Studies in Advanced Mathematics

T0136047

Several Complex Variables and the Geometry of Real Hypersurfaces

Studies in Advanced Mathematics

JOHN P. D'ANGELO
University of Illinois, Department of Mathematics

Several Complex Variables and the Geometry of Real Hypersurfaces

CRC Press
Taylor & Francis Group
Boca Raton London New York

CRC Press is an imprint of the
Taylor & Francis Group, an **informa** business

CRC Press
Taylor & Francis Group
6000 Broken Sound Parkway NW, Suite 300
Boca Raton, FL 33487-2742

First issued in paperback 2019

© 1993 by Taylor & Francis Group, LLC
CRC Press is an imprint of Taylor & Francis Group, an Informa business

No claim to original U.S. Government works

ISBN-13: 978-0-8493-8272-7 (hbk)
ISBN-13: 978-0-367-40248-8 (pbk)

Visit the Taylor & Francis Web site at
http://www.taylorandfrancis.com

and the CRC Press Web site at
http://www.crcpress.com

To the memory of my mother,
Ethel G. D'Angelo.

Contents

Preface

This book discusses some geometric questions in the theory of functions of several complex variables. It describes some beautiful connections among algebra, geometry, and complex analysis. The unifying theme will be the interplay between real varieties and complex varieties. The connection between these two classes of varieties arises often in the following setting. Suppose that Ω is a domain in $\mathbb{C}^n$ with smooth boundary M. As a real hypersurface in a complex manifold, M inherits a rich geometric structure. There is a certain amount of complex structure on its tangent spaces; consideration of this structure leads us to the Levi form and other properties that belong to CR geometry. For many analytic problems, however, especially those that involve weakly pseudoconvex domains, one needs higher order invariants. It turns out that these ideas achieve their clearest exposition by algebraic geometric methods rather than differential geometric ones. The algebraic geometric methods appear when one studies whether the real object M contains any complex analytic varieties, or more generally, how closely ambient complex analytic varieties can contact M. We now give two examples of this phenomenon.

It is a classical result dating back to Poincaré that there is no biholomorphism from the unit polydisc to the unit ball in dimensions two or more. Poincaré proved this by computing the automorphism groups of the two domains. Following an approach of Remmert and Stein [RS], one proves easily that there is no proper holomorphic mapping from a product of bounded domains to another bounded domain, unless the image domain contains complex analytic sets in its boundary. Thus it is the absence of complex varieties in the sphere, rather than the strong pseudoconvexity of the sphere, that governs this aspect of the function theory.

Next, consider the geometry associated with the theory of the $\bar{\partial}$-Neumann problem. By 1962, J. J. Kohn [K1] had already solved this problem for smoothly bounded strongly pseudoconvex domains. Kohn's solution used the method of L^2 estimates for solutions of the Cauchy–Riemann equations. A particular kind of *a priori* estimate that arose in this context became known as a subelliptic estimate. Such an estimate has many consequences for function theory, such as local regularity of the canonical solution to the inhomogeneous Cauchy–

Riemann equations and regularity for the Bergman projection. Kohn naturally asked for necessary and sufficient conditions on a pseudoconvex domain for the validity of these estimates. After considerable work in this area, Catlin [C3] solved the problem by proving that a necessary and sufficient condition for a subelliptic estimate at a boundary point p is the boundedness of the order of contact of all ambient complex analytic varieties with the boundary at p. Such a point is a "point of finite type"; the geometry of points of finite type is a significant part of this book.

These examples suggest that a thorough study of the geometry of the real boundary, by way of its contact with ambient complex analytic objects, is a significant aspect of the theory of functions of several complex variables. An important principle appears in each of the preceding examples. The appropriate generalization of strong pseudoconvexity for a hypersurface is the absence of complex varieties in the hypersurface, or more precisely, the finiteness of the order of contact of all such varieties with the hypersurface. This amounts to a finite-order degeneracy in the Levi form, so it is possible to apply algebraic ideas. The reasoning is analogous to that used in passing from the implicit function theorem to the Weierstrass preparation theorem, or in generalizing from local biholomorphisms to mappings that are finite to one. Such considerations arise in many analytic problems; we allude to such problems throughout the book.

The first two chapters are background material. In Chapter 1 we begin with the necessary preliminaries on the theory of analytic functions of several complex variables. We give a complete proof of the Remmert–Stein theorem to motivate the basic theme of the book. How does one decide whether a real hypersurface in $\mathbb{C}^n$ contains complex analytic varieties, and why does one care? The introductory chapter also includes some discussion of implicit functions, the Weierstrass preparation and division theorems, and their consequences. We also prove the Nullstellensatz for holomorphic germs. This theorem is fundamental to the approach taken in the rest of the book. This material enables the book to be used for a graduate course.

In Chapter 2 we recall enough commutative algebra to understand the rudiments of the theory of intersection numbers. The main point is to assign a positive number (multiplicity) to an ideal that defines a zero-dimensional variety. We give many ways to measure this singularity: topological, geometric, and algebraic. I hope to help the reader become completely familiar with the properties of such "finite mappings." The theory is particularly elegant in case the number of generators of the ideal equals the dimension of the underlying space. Since this theory does not seem to be well known to analysts, we give many equivalent definitions and reformulations of such a multiplicity. There are several other measurements; we give inequalities relating them. Of basic importance is the relationship between an invariant called the "order of contact" of an ideal and the usual intersection number, or codimension, of an ideal. The order of contact makes sense for ideals in other rings and thus applies to real hypersurfaces as well. This chapter also includes many concrete computations of various intersection numbers. Several important algebraic facts are stated but not proved in this chapter, although references are provided.

In Chapter 3 we discuss the local geometry of real hypersurfaces. We begin with a differential geometric discussion and then proceed to a more algebraic geometric one. We determine precisely which algebraic and real analytic hypersurfaces contain positive dimensional complex analytic varieties. To do so, we associate a family of ideals of holomorphic functions with each point on such a hypersurface. This reduces the geometry to the algebraic ideas of Chapter 2.

In Chapter 4 we merge the ideas of Chapters 2 and 3. We discuss carefully the family of ideals associated with each point of a smooth hypersurface. We express the maximum order of contact of complex analytic varieties with a smooth hypersurface by applying the order of contact invariant to these ideals of holomorphic functions. The theorems here include the local boundedness of the maximum order of contact of complex analytic varieties with a real hypersurface, sharp bounds in the pseudoconvex case, and the result that a point in a real analytic hypersurface is either a point of finite type or lies in a complex analytic variety actually lying in the hypersurface. This last result is a kind of uniform boundedness principle.

In Chapter 5 we return to proper holomorphic mappings. We emphasize mappings between balls in different dimensions, thereby seeing more applications of the algebraic reasoning that dominates this book. Our interest here is in the classification of proper mappings between balls; a recent theorem of Forstnerič shows that sufficiently smooth proper mappings must be rational. Restricting ourselves to this case affords us the opportunity to be concrete; in particular, there are many interesting explicit polynomial mappings of balls. In this chapter we give a classification and thorough discussion of the polynomial proper mappings between balls in any dimensions. There is also a new result on invariance of proper mappings under fixed-point–free finite unitary groups. We determine precisely for which such groups Γ there is a Γ-invariant rational proper holomorphic mapping between balls. The methods here are mostly elementary, although we use without proof two important ingredients. One is the theorem of Forstnerič mentioned above. The other is the classification of fixed-point–free finite unitary groups from Wolf's book on spaces of constant curvature [Wo]. I view the chapter as an enjoyable hiatus before we tackle Chapter 6. The rest of the book requires none of the results from Chapter 5, so the reader interested in subelliptic estimates may omit it. I included this chapter because it helps to unify some of the underlying ideas of complex geometry. The treatment of z and its conjugate $\bar{z}$ as independent variables epitomizes this approach; we see several surprising applications in the study of proper mappings between balls. The methods here may apply to other problems; it seems particularly valuable to have a notion of Blaschke product in several variables.

In the final two chapters we discuss the relationship between the first five chapters and the geometry of the $\bar{\partial}$-Neumann problem. Many of the geometric questions discussed in this book were motivated by the search for geometric conditions on a pseudoconvex domain that guarantee subelliptic estimates for the $\bar{\partial}$-Neumann problem. Chapter 6 is deep and difficult. First we recall some of the existence and regularity results of Kohn for the $\bar{\partial}$-Neumann problem. Assuming some formal properties of tangential pseudodifferential operators, we discuss Kohn's method of subelliptic multipliers. This introduces the reader to methods of partial differential equations in complex analysis. Following Kohn in the real analytic pseudoconvex case, we prove that the subelliptic multipliers form a radical ideal and discuss their relation to real varieties of positive holomorphic dimension. We include the Diederich–Fornaess theorem [DF1] about the relation between such varieties and complex analytic ones. Unfortunately, I was unable to include a proof of Catlin's theorem that subellipticity is equivalent to "finite type" in the smooth case, although a reader who gets this far will be able to understand Catlin's work.

In Chapter 7 we discuss the methods of Bell [Be1]; this approach uses properties of the Bergman projection to prove boundary smoothness of proper holomorphic mappings between pseudoconvex domains that satisfy condition R. Condition R is a regularity condition for the Bergman projection; it holds whenever there is a subelliptic estimate. The results on boundary smoothness justify the geometric considerations of the earlier chapters, for they imply that boundary invariants of smoothly bounded pseudoconvex domains of finite type are biholomorphic invariants. We give some applications and include the statements of some results on extensions of CR functions. The reader can consult Boggess's book [Bg] for the latest results here. Thus we glimpse deep analytic results proved since 1970 and discover some directions for future research.

The book closes with two lists of problems. The first list consists of mostly routine exercises; their primary purpose is for the reader to check whether he follows the notation. The second list consists of very difficult problems; many of these are open as of January 1992.

I believe that combined use of the algebraic and analytic ideas herein forms a useful tool for attacking several of the field's open problems. In particular consider the following analogy. Strongly pseudoconvex points correspond to the maximal ideal, while points of finite type correspond to ideals primary to the maximal ideal. Making sense of this simple heuristic idea is perhaps the raison d'être for the writing of this book.

The prerequisites for reading the book are roughly as follows. The reader must certainly know the standard facts about analytic functions of one complex variable, although few deep theorems from one variable are actually used in proofs. The reader must know quite a bit of advanced calculus, what a smooth manifold is, what vector fields and differential forms are, how to use Stokes' theorem, the inverse and implicit function theorems, and related ideas. The reader also needs to know some algebra, particularly some of the theory of commutative rings. We derive many algebraic properties of the ring of convergent power series, but state and use others without proof. Our proof of the Nullstellensatz uses some algebra. We assume also a formula from intersection theory that relies on lengths of modules and the Jordan–Holder theorem. We do not prove this formula, but we perform detailed computations with it to aid our understanding.

Chapters 3 and 4 rely almost entirely on the ideas of Chapters 1 and 2. Most of Chapter 5 is elementary; the deeper things needed are discussed in the summary above. Chapters 6 and 7 require more. A reader should know the basics of linear partial differential equations, including pseudodifferential operators and Sobolev spaces. Some facts about real analytic sets, such as the Lojasiewicz inequality, also appear in Chapter 6.

The book has many computations and examples as well as definitions and theorems. There are some incomplete proofs. I have not written a completely formal book; at times I substitute examples or special cases for rigorous proofs. The reader can glimpse deep aspects of the subject, but may have to work hard to master them. I hope that the point of view I have taken will help and the compromises I have made will not hurt.

I apologize for any omissions or errors. To some extent they reflect my taste and to some extent they expose my ignorance. Mathematics is both an individual and a team game. I have tried to cast my work in a broader context; the inherent risk in such a plan is that I must discuss many things better understood by others. I hope that I have been faithful to the subject in these discussions.

I am indebted to many. First I thank J. J. Kohn for introducing me to the problem of finding geometric conditions for subelliptic estimates, and for many conversations over the years. Next I thank David Catlin for many lively discussions and arguments about the geometric considerations involved; Catlin's theorem affords additional justification for the geometry we study here. Additional thanks are due Salah Baouendi, Steve Bell, Jim Faran, John Fornaess, Franc Forstnerič, Bob Gunning, Steve Krantz, Laszlo Lempert, Michael Range, and Linda Rothschild for their research in this area and for many helpful discussions about complex analysis. I acknowledge especially Steve Krantz for inviting me to Washington University for the spring of 1990 and giving me the opportunity and encouragement needed to write this book. Dan Grayson commented helpfully on a very preliminary version of Chapter 2. John Fornaess, Jeff McNeal, and Emil Straube reviewed a preliminary version of the manuscript and each made some valuable remarks. Many other mathematicians have contributed to the development of this subject or have aided me in my study of complex analysis. A list of such people would make the preface longer than the book; I am indebted to all. I thank in particular my colleagues and students at both the University of Illinois and Washington University for listening to lectures on this material. The NSF, the Institute for Advanced Study, and the Mittag-Leffler Institute all provided support at various times in my career.

I thank my family, friends, and basketball teammates for providing me many happy distractions, and I dedicate this book to the memory of my mother.

1

Holomorphic Functions and Mappings

1.1 Preliminaries

1.1.1 Complex *n*-dimensional space

We denote by $\mathbb{C}^n$ the complex vector space of dimension n with its usual Euclidean topology. A domain in $\mathbb{C}^n$ is an open, connected set. The collection of open polydiscs constitute a basis for the open subsets of $\mathbb{C}^n$, where a polydisc is a product of discs. Let $z = \left(z^1, \ldots, z^n\right)$ denote coordinates on $\mathbb{C}^n$. Our notation for the polydisc of multi-radius $r = \left(r^1, \ldots, r^n\right)$ centered at w will be

$$P_r\left(w\right) = \left\{z : \left|z^j - w^j\right| < r^j \quad j = 1, \ldots, n\right\}. \tag{1}$$

An alternate basis for the topology is the collection of open balls. The ball of radius r about w is defined by

$$B_r\left(w\right) = \left\{z : \left\|z - w\right\|^2 < r^2\right\} \tag{2}$$

where $\|z - w\|^2 = \sum_{j=1}^n |z^j - w^j|^2$ denotes the squared Euclidean distance. Notice that we use the same symbol for the radius of a ball and the multi-radius of a polydisc. One of our first tasks will be to show that the polydisc and the ball have very different function theories.

The geometry of $\mathbb{C}^n$ reveals immediately the importance of the interaction between the real and the complex. Consider the underlying real variables defined by

$$z^j = x^j + iy^j = \mathrm{Re}\left(z^j\right) + i\,\mathrm{Im}\left(z^j\right). \tag{3}$$

For a basis of the complex-valued differential one-forms on $\mathbb{C}^n$, we can take either the underlying real differentials, or the complex differentials $\{dz^1, \ldots, dz^n,$

$d\bar{z}^1, \ldots, d\bar{z}^n\}$. These are related by the formulas

$$dx^j = \frac{1}{2}\left(dz^j + d\bar{z}^j\right)$$

$$dy^j = \frac{1}{2i}\left(dz^j - d\bar{z}^j\right). \tag{4}$$

The differential of a smooth function f can be written therefore in two ways:

$$df = \sum_{j=1}^{n} \frac{\partial f}{\partial x^j} dx^j + \frac{\partial f}{\partial y^j} dy^j \tag{5}$$

$$df = \partial f + \bar{\partial} f = \sum_{j=1}^{n} \frac{\partial f}{\partial z^j} dz^j + \sum_{j=1}^{n} \frac{\partial f}{\partial \bar{z}^j} d\bar{z}^j. \tag{6}$$

It is important to observe that there is no choice in the definition of the partial derivative operators

$$\frac{\partial}{\partial z^j} = \frac{1}{2}\left(\frac{\partial}{\partial x^j} - i\frac{\partial}{\partial y^j}\right)$$

$$\frac{\partial}{\partial \bar{z}^j} = \frac{1}{2}\left(\frac{\partial}{\partial x^j} + i\frac{\partial}{\partial y^j}\right). \tag{7}$$

These definitions follow immediately from equating the two formulas for the differential df and plugging in the formulas for the real differentials in terms of the complex differentials.

In this book we will work only with the complex differentials. Differential one-forms that are (functional) combinations of the dz^j are called forms of type (1,0), and those that are combinations of the $d\bar{z}^j$ are called forms of type (0,1). The coefficient functions are in general smooth complex-valued functions, although in Chapter 6 the coefficient functions will be more general. We consider also differential forms of type (p, q). It is standard to write $dz^J = dz^{j_1} \wedge dz^{j_2} \ldots \wedge dz^{j_q}$ for the basic $(q, 0)$ forms. For complex vector fields one uses similar terminology; thus a (functional) combination of the $\partial/\partial z^j$ is called a vector field of type (1,0), and a combination of the $\partial/\partial \bar{z}^j$ is called a vector field of type (0,1). See [Ra] for a thorough discussion of these topics.

Perhaps the most fundamental object in complex analysis is the Cauchy–Riemann operator, whose definition we now give. Notice in Definition 1 that $d\bar{z}^j$ is a differential one-form while dz^I and $d\bar{z}^J$ denote forms of unspecified degree.

DEFINITION 1 *The Cauchy–Riemann operator $\bar{\partial}$ is defined on continuously differentiable functions by*

$$\bar{\partial} f = \sum_{j=1}^{n} \frac{\partial f}{\partial \bar{z}^j} d\bar{z}^j \tag{8}$$

and on differential forms of higher degree by

$$\overline{\partial}\left(\sum_{I,J} a_{IJ}\, dz^I \wedge d\bar{z}^J\right) = \sum_{I,J} \overline{\partial} a_{IJ} \wedge dz^I \wedge d\bar{z}^J$$

$$= \sum_{I,J} \sum_{j=1}^{n} \frac{\partial a_{IJ}}{\partial \bar{z}^j}\, d\bar{z}^j \wedge dz^I \wedge d\bar{z}^J. \tag{9}$$

Note that $\overline{\partial} f$ is defined to be the $(0,1)$ part of df. The extension of the definition of $\overline{\partial}$ to forms of all degrees yields in the usual fashion an operator satisfying $\overline{\partial}\overline{\partial} = 0$. The resulting complex is called either the Cauchy–Riemann complex or the Dolbeault complex.

1.1.2 Holomorphic functions

The classical idea from complex analysis in one variable—that a complex analytic function is one that depends on z, but not on its conjugate $\bar{z}$—manifests itself in the Cauchy–Riemann equations. This heuristic idea arises often in this book. By now it is standard to define holomorphic functions as we do here.

DEFINITION 2 *Let Ω be a domain in $\mathbb{C}^n$ and suppose that $f : \Omega \to \mathbb{C}$ is a continuously differentiable function. Then f is called holomorphic, or complex analytic, on Ω if*

$$\overline{\partial} f = 0 \tag{10}$$

or equivalently if

$$\frac{\partial f}{\partial \bar{z}^k} = 0 \qquad k = 1, \dots, n \tag{11}$$

at each point of Ω.

This first-order system of linear partial differential equations is called the (homogeneous) Cauchy–Riemann equations. One of the great advances in the subject was the idea that one could study complex analysis by treating the inhomogeneous Cauchy–Riemann equations $\overline{\partial} f = g$, where g is a differential form of type $(0,1)$ satisfying the necessary compatibility equations $\overline{\partial} g = 0$, as a problem in partial differential equations. The applications of this sort of reasoning have dominated the subject in the last thirty years. See Chapter 6 for more information on the Cauchy–Riemann equations.

For completeness we recall some of the basic theorems about holomorphic functions. Many of the proofs follow from iterating the Cauchy integral formula from the one-variable case. Let us introduce some notation before stating these results. When P is a polydisc, we let $\partial_o P$ denote the distinguished boundary of P, namely the product of the boundaries of each disc. Multi-index notation is common in this book; it is easier to read and frequently it better represents the ideas than more complicated notation. Two standard uses of multi-index notation are that

$$dw = dw^1 \wedge \cdots \wedge dw^n \tag{12}$$

and

$$\frac{1}{w - z} = \frac{1}{\left(w^1 - z^1\right) \cdots \left(w^n - z^n\right)}. \tag{13}$$

The useful abbreviation that

$$\frac{1}{(w - z)^{\alpha+1}} = \prod_{j=1}^{n} \frac{1}{\left(w^j - z^j\right)^{\alpha_j+1}}$$

follows from (13).

It is immediate from the definition that a holomorphic function of several variables is holomorphic in each variable separately. This means that the function of one variable that results by fixing the values of all but one of the variables is holomorphic where it is defined. The converse is a deep, albeit seldom used, theorem of Hartogs; see [Hm] for a proof. See also [Shi] for recent generalizations.

THEOREM 1

A complex-valued function on a domain in $\mathbb{C}^n$ is holomorphic if and only if it is holomorphic in each variable separately.

It is easy to prove the weaker statement that a continuous function that is holomorphic in each variable separately is holomorphic. This follows from the proof of the Cauchy integral formula for polydiscs. Since we do not need any results on separate holomorphicity, we state the integral formula for holomorphic functions. The simple proof given applies under the hypotheses of "continuous and holomorphic in each variable separately" on the function f.

THEOREM 2

(Cauchy integral formula for polydiscs). *Suppose that P is an open polydisc, and that f is holomorphic on P and continuous on the closure of P. Then, for $z \in P$,*

$$f(z) = \frac{1}{(2\pi i)^n} \int_{\partial_o P} \frac{f(w)}{w - z}\, dw. \tag{14}$$

PROOF   Since f is holomorphic, it is holomorphic as a function of its first variable and can be represented as a Cauchy integral. The resulting integrand is holomorphic in the rest of the variables. The result now follows from iterating the Cauchy integral formula in one variable, and finally using the continuity of f to replace the iterated integral with an integral over the n-cycle $\partial_o P$. ∎

Two fundamental consequences of this version of the Cauchy integral formula are the convergence properties for sequences of holomorphic functions and the power series representation for a holomorphic function. We devote the next section to convergence properties of holomorphic functions.

1.1.3   The topological vector space $O(\Omega)$

Let us denote by $O(\Omega)$ the algebra of all holomorphic functions on Ω. See [GR, p. 2] for a discussion of the etymology of this standard notation. It is natural to conceive of $O(\Omega)$ as a subspace of C(Ω), where C(Ω) denotes the continuous functions on Ω. Thus one makes $O(\Omega)$ into a topological vector space by use of the strict inductive limit topology. This means that a sequence $\{f_k\}$ converges in $O(\Omega)$ if it converges uniformly on each compact subset of Ω. The result is that $O(\Omega)$ is a complete, metrizable, locally convex, topological vector space. Such a space is called a Fréchet space. One sees that the space is metrizable by choosing an exhausting sequence of compact subsets. As is customary, we write $L \subset\subset \Omega$ to denote that the closure of the subset L in Ω is compact. When $K \subset\subset \Omega$ is any compact subset, and $f \in O(\Omega)$, we write

$$\|f\|_K = \sup_{z \in K} |f(z)| \tag{15}$$

for the corresponding seminorm. The classical fact that the uniform limit on compact subsets of analytic functions is itself analytic can be stated more succinctly by saying that $O(\Omega)$ is a closed subspace of C(Ω). We prove this and also some of the standard compactness results about $O(\Omega)$. First let us recall the Arzela–Ascoli theorem and the concept of equicontinuity. A family S of continuous functions on a compact set K is equicontinuous if

$$\forall \epsilon > 0, \exists \delta > 0 : \quad \forall f \in S, |f(z) - f(w)| < \epsilon$$
$$\text{whenever } \|z - w\| < \delta. \tag{16}$$

This theorem relates equicontinuity to compactness; in complex analysis, bounded sets of holomorphic functions are automatically equicontinuous, so the theorem becomes very powerful. See [Ah] or any other good complex analysis text for more information.

THEOREM 3

(Arzela–Ascoli). *Let K be a compact set, and let* C(K) *be the Banach space of complex-valued continuous functions on K in the supremum norm. Suppose* *that S is a subset of $C(K)$. Then for S to be compact it is necessary and sufficient that S satisfy all of the following three conditions:*

1. *S is closed.*

2. *S is equicontinuous.*

3. *S is bounded, i.e., $\{\|f\|_K : f \in S\}$ is a bounded subset of $\mathbf{R}$.*

We will use Theorem 3 momentarily. In order to see that $O(\Omega)$ is closed in $C(\Omega)$ and to apply compactness arguments, it is necessary to first recall the Cauchy estimates.

THEOREM 4

Suppose that Ω is an open subset of $\mathbf{C}^n$, that $f \in O(\Omega)$, and that $K \subset\subset \Omega$ is any compact subset. Suppose also that $z \in \Omega$, and that the closure of the polydisc $P_r(z)$ lies within Ω. The following inequalities hold:

1. *For any multi-index α,*

$$|D^\alpha f(z)| \le \frac{\alpha!}{r^\alpha} \|f\|_{\partial_o P_r(z)}. \tag{17}$$

2. *For any multi-index α there is a constant c_α depending on α, K, Ω but not on f, so that*

$$\|D^\alpha f\|_K \le c_\alpha \|f\|_\Omega. \tag{18}$$

PROOF   First we suppose that z is any point in Ω. Choose a multi-radius r such that the closure of $P_r(z)$ is contained in Ω. One applies the operator D^α to both sides of the Cauchy integral formula, to obtain the formula that

$$D^\alpha f(z) = \left(\frac{1}{2\pi i}\right)^n \int\limits_{\partial_o P} \frac{\alpha!}{(w-z)^{\alpha+1}} f(w)\, dw. \tag{19}$$

An obvious estimate of the integral yields the inequalities that

$$|D^\alpha f(z)| \le \frac{\alpha!}{r^\alpha} \|f\|_{\partial_o P_r(z)} \le \frac{\alpha!}{r^\alpha} \|f\|_\Omega. \tag{20}$$

This proves the first statement. To prove the second statement, we wish to take the supremum over $z \in K$. Since K is compact, the distances from each $z \in K$ to $b\Omega$ are bounded away from zero. Thus the reciprocals of these distances are bounded. By choosing a "large" polydisc corresponding to each point, we may therefore assume that the expressions $1/r^\alpha$ in (20) are uniformly bounded as z varies throughout K. Taking then the supremum over z in inequality (20) yields the desired conclusion that

$$\|D^\alpha f\|_K = \sup_K \|D^\alpha f(z)\| \le c_\alpha \|f\|_\Omega. \quad \blacksquare \tag{21}$$

THEOREM 5
The space $O(\Omega)$ of holomorphic functions on Ω is a closed subspace of $C(\Omega)$ in the strict inductive limit topology.

PROOF It is trivial that the holomorphic functions form a subspace. To show that this subspace is closed, suppose that $\{f_k\}$ is a sequence in $O(\Omega)$ that converges uniformly on each compact subset of Ω to the continuous limit f. Suppose that K is any compact subset of Ω. Consider the sequence $\{Df_k\}$ defined by any first-order derivative. By the Cauchy estimates, there is a constant c_K so that

$$\|D(f_n - f_m)\|_K \leq c_K \|f_n - f_m\|_K. \tag{22}$$

Thus the sequence of first derivatives converges uniformly on each compact subset, so the limit function f is continuously differentiable and $\overline{\partial}f = \lim\left(\overline{\partial}f_k\right) = 0$. Thus f is holomorphic, so $O(\Omega)$ is closed in $C(\Omega)$. ∎

Another standard and important limiting argument is given by the next theorem.

THEOREM 6
Suppose that $\{f_j\}$ is a sequence of holomorphic functions on Ω and that $\|f_j\|_K$ is bounded (independently of j) for each compact subset K of Ω. Then there is a subsequence $\{f_{j_k}\}$ that converges in $O(\Omega)$.

PROOF Choose an exhausting sequence of compact subsets K_m of Ω. If we can show that the sequence $\{f_j\}$ is equicontinuous on each K_m, then it would follow from the Arzela–Ascoli theorem that this sequence would have compact closure. There would then be, for each m, a subsequence $\{f_{j_{m(k)}}\}$ convergent on K_m. By the Cantor diagonalization process we could then extract one subsequence that is convergent on every K_m. Thus it is enough to establish equicontinuity on each compact subset. For any holomorphic f, and any compact subset K, it follows from the mean value theorem and the Cauchy estimates that

$$|f(z) - f(w)| \leq \|z - w\| \ \|Df\|_K$$
$$\leq c\|z - w\| \ \|f\|_K. \tag{23}$$

If the sequence $\{f_j\}$ is also bounded, then for each K, $\{\|f_j\|_K\}$ is a bounded set of numbers, and we see that

$$|f_j(z) - f_j(w)| \leq c_K \|z - w\|. \tag{24}$$

This establishes the equicontinuity on K and completes the proof. ∎

There is a possibility of confusion about bounded sets. The definition of boundedness in a Fréchet space such as $O(\Omega)$ is as follows: A subset S is bounded if and only if, for every compact set K,

$$\{\|f\|_K : f \in S\} \tag{25}$$

is a bounded set of numbers. Note that this is equivalent to any reasonable notion of boundedness for a subset of a topological vector space; if, however, one makes $O(\Omega)$ into a metric space by using the seminorms coming from an exhausting sequence of compact subsets, then the resulting notion of boundedness differs. The concept of boundedness in the metric space sense for a subset of $O(\Omega)$ does not arise in this book.

COROLLARY 1
A subset of $O(\Omega)$ is compact if and only if it is closed and bounded.

All these arguments apply just as well for holomorphic mappings $f : \Omega \to \mathbb{C}^n$. It is convenient to use the notation $O(\Omega_1, \Omega_2)$ for the set of holomorphic mappings from Ω_1 into Ω_2, when these are open domains in any dimensions.

1.1.4 Elementary properties of holomorphic functions

We show in this section that a holomorphic function has a local convergent power series and conversely that a convergent power series represents a holomorphic function. Power series methods are basic in complex analysis, and they are particularly valuable in this book because of their relationship to commutative algebra.

THEOREM 7
A continuously differentiable function f is holomorphic on Ω if and only if, for all p in Ω, there is a neighborhood $U = U_p$ such that f can be represented as a multiple power series

$$f(z) = \sum_{|a|=0}^{\infty} c_a (z - p)^a \tag{26}$$

where the series converges absolutely and uniformly on each compact subset of U_p. One can choose U_p to be any polydisc that is centered at p and also lies within Ω.

PROOF   The proof is essentially the same as in one variable, so we give only a sketch. Assuming that f is holomorphic, one begins with the Cauchy integral formula and writes

$$\frac{1}{w - z} = \frac{1}{w - p - (z - p)} = \frac{1}{(w - p)\left(1 - \frac{z-p}{w-p}\right)} \tag{27}$$

and expands in a multiple geometric series. This series converges absolutely and uniformly on compact subsets of the polydisc

$$U = \left\{ z : \frac{|z^j - p^j|}{|w^j - p^j|} < 1, \qquad j = 1, \ldots, n \right\}. \tag{28}$$

An interchange of integration and summation yields the result, and the formula

$$c_a = \left(\frac{1}{2\pi i} \right)^n \int_{\partial_o P} \frac{f(w)}{(w - p)^{a+1}} \, dw. \tag{29}$$

To prove the converse assertion, note that each partial sum of the power series is a polynomial in z and hence is holomorphic. By Theorem 5, the uniform limit on compact subsets of holomorphic functions is holomorphic. Thus a convergent power series in z represents a holomorphic function. ∎

COROLLARY 2
A holomorphic function is infinitely differentiable.

The Cauchy integral formula for polydiscs is important for obtaining many elementary properties of holomorphic functions. Other properties follow automatically from the one-variable case. We leave it to the reader to verify the following statements.

THEOREM 8
A nonconstant holomorphic function is an open mapping.

Note, however, that nonconstant holomorphic mappings (that is, $\mathbb{C}^n$-valued holomorphic functions) need not be open mappings. See Chapter 2 for more information about this topic.

THEOREM 9
(The maximum principle). *If f is holomorphic on a domain Ω, and $|f|$ achieves its maximum at an interior point of Ω, then f is constant.*

THEOREM 10
A holomorphic function on a domain that has a zero of infinite order at one point vanishes identically.

1.1.5 Polarization of analytic identities

We conclude this preliminary section with a simple but powerful proposition that involves treating z and $\bar{z}$ as independent variables. Since these functions have linearly independent differentials, they should correspond to "independent" functions. The idea of treating them in this manner has surprising consequences, so it is valuable to indicate precisely what this means.

PROPOSITION 1

Suppose that Ω is a domain in $\mathbb{C}^n$ and let $\Omega^* = \{z : \bar{z} \in \Omega\}$ be its conjugate domain. Suppose that

$$H : \Omega \times \Omega^* \to \mathbb{C} \tag{30}$$

is a holomorphic function of the 2n complex variables (z, w), and that

$$H(z, \bar{z}) = 0 \tag{31}$$

for all z in Ω. Then

$$H(z, w) = 0 \tag{32}$$

for all $(z, w) \in \Omega \times \Omega^*$.

Replacing $\bar{z}$ in an analytic identity with a new variable, while keeping z the same, is called polarization. The technique of polarization is especially important in the theory of mappings between real analytic hypersurfaces. It is the basis of reflection principles and extension results [Fo2,We2].

PROOF OF PROPOSITION 1   There are many possible proofs; one amounts to showing that H vanishes on a copy of real Euclidean space of dimension $2n$. We present here instead an iconoclast's proof. Without loss of generality, we may assume that Ω is a ball about 0. Expand H in a power series in $(z, w) \in \Omega \times \Omega^*$ and introduce polar coordinates in each variable. In multi-index notation,

$$z = re^{i\theta} \tag{33}$$

and hence,

$$z^a = r^a e^{ia\theta} = \prod_{j=1}^{n} r_j^{a_j} e^{ia_j\theta_j}. \tag{34}$$

Substitution of this into the power series representation for H yields

$$0 = \sum_a \sum_b c_{ab} \, r^{a+b} e^{i\theta(a-b)}. \tag{35}$$

Replacing $a - b$ with k, and equating Fourier coefficients, we see that

$$0 = \sum_b c_{(b+k)b} r^{2b} \qquad \forall k \in Z^n. \tag{36}$$

Equating coefficients to 0 in each of these convergent power series tells us that each coefficient c_{ab} vanishes. Thus H vanishes identically. ∎

A particular application of this sort of reasoning belongs in a first course in complex variables. Suppose that $u = u(x, y)$ is a harmonic function and one wishes to find a holomorphic function $f = f(z)$ for which u is the real part. Suppose without loss of generality that

$$f(0) = u(0, 0) = 0. \tag{37}$$

Then there is a simple formula for f, namely

$$f(z) = 2u\left(\frac{z}{2}, \frac{z}{2i}\right). \tag{38}$$

Ahlfors [Ah] gives this formula without a rigorous proof. Here is a simple proof, based on Proposition 1.

Since u is harmonic, it is real analytic. (This means that it has a local convergent power series representation in the underlying real variables.) It is hence valid to substitute

$$x = \frac{z + \bar{z}}{2}$$
$$y = \frac{z - \bar{z}}{2i} \tag{39}$$

into the formula for $u(x, y)$. The result is the identity that

$$f(z) + \overline{f(z)} = 2u\left(\frac{z + \bar{z}}{2}, \frac{z - \bar{z}}{2i}\right). \tag{40}$$

It is then necessary to extract the holomorphic terms. According to Proposition 1, it is legitimate to substitute $\bar{z} = 0$ into this identity. The result follows, assuming (as we have done) that $f(0) = 0$. ∎

1.2 Holomorphic mappings

1.2.1 Biholomorphic and proper mappings

It is of great interest in mathematics to decide in a given context whether distinct objects are essentially the same. Suppose that Ω_1, Ω_2 are domains in $\mathbb{C}^n$. These domains will have the same function theory if there is a bijective holomorphic mapping

$$f : \Omega_1 \to \Omega_2 \tag{41}$$

with a holomorphic inverse. Such a mapping is called a biholomorphism, or biholomorphic mapping, and the domains are said to be biholomorphically equivalent. More general than a biholomorphic mapping is a proper holomorphic mapping. A mapping

$$f : \Omega_1 \to \Omega_2 \tag{42}$$

is called proper if whenever K is a compact subset of Ω_2, $f^{-1}(K)$ is a compact subset of Ω_1. The inverse of a biholomorphic mapping is holomorphic, hence continuous, and therefore takes compact sets to compact sets; thus a biholomorphic mapping is necessarily proper. The simplest examples of proper mappings that are not biholomorphic are the mappings

$$z \to z^m \tag{43}$$

for $m > 1$. These are proper from the disc to itself or from $\mathbb{C}$ to itself. There are analogous mappings in several variables; they must necessarily map balls into higher dimensional balls. We study these in Chapter 5.

In complex analysis in several variables, it is seldom the case that given domains are biholomorphically equivalent, so one also wishes to know whether there is a proper holomorphic mapping from one domain onto another. There are theorems in function theory of several complex variables stating that proper mappings between certain classes of domains (in the same dimension) are necessarily biholomorphic (see [Bed]). In the equidimensional case it is seldom true that there is a proper holomorphic mapping between two given domains, although it becomes easier when the target domain is of larger dimension. There is a brief discussion of these ideas in Chapter 5. In Chapter 7 we also discuss some general results to the effect that proper mappings between domains (in the same dimension) with pseudoconvex smooth boundaries that satisfy finite type conditions (of the sort studied in this book) must extend smoothly to the boundaries. In such a context it becomes meaningful to assign boundary invariants to domains. For a proper mapping between two given domains to exist, it is necessary that these boundary invariants transform appropriately. Unless these boundary invariants match up correctly, there can be no proper mapping between these domains. The simplest case of this sort of phenomenon is that of the polydisc and the ball. There is no proper mapping from a polydisc to a ball because the polydisc has complex analytic structure in its boundary and the ball does not. This is the subject of the next section.

1.2.2 Inequivalence of the polydisc and the ball

By the Riemann mapping theorem, every simply connected proper subdomain of $\mathbb{C}$ is biholomorphically equivalent to the unit disc. Thus the disc serves as a model domain in one dimension. The situation is considerably more complicated in higher dimensions. It is in general impossible to find even a proper holomorphic mapping between domains that are diffeomorphic. Thus there are many model domains; in particular, the polydisc and the ball have quite different function theories.

Poincaré showed in 1907 that the unit polydisc and the unit ball are not biholomorphically equivalent. He did so by computing the automorphism groups of both domains; as real Lie groups they are of different dimensions. Today there are many proofs of this inequivalence. We prove, essentially following

Remmert and Stein, a more general result in this section, which shows already the value in understanding whether the boundary $b\Omega$ of a domain Ω contains complex analytic sets.

We begin with a simple but useful characterization of continuous proper mappings between bounded domains. Note, as a consequence: a mapping that has a continuous extension to the boundary must map the boundary of the domain to the boundary of the target.

LEMMA 1

A continuous mapping $f : \Omega_1 \to \Omega_2$ between bounded domains is proper if and only if, for every sequence $\{z_k\}$ tending to $b\Omega_1$, the sequence $\{f(z_k)\}$ tends to $b\Omega_2$.

PROOF If the sequence condition fails, then there is a subsequence $\{f(z_{k_j})\}$ that is convergent to a point in the interior of Ω_2. These values and the limit form a compact subset of Ω_2. The inverse image of this sequence is a subsequence of $\{z_k\}$ and thus tends to $b\Omega_1$. Thus the inverse image of a compact subset is not compact, so the mapping is not proper. Conversely, suppose that K is compact in Ω_2. The inverse image $f^{-1}(K)$ is closed by continuity, so to verify compactness it remains only to show that it is bounded away from $b\Omega_1$. If not, then we can find a sequence $\{z_k\}$ in $f^{-1}(K)$ that tends to $b\Omega_1$. We are assuming that its image lies within the compact set K and thus is bounded. This means the sequence property fails. Therefore, if the sequence property holds, then $f^{-1}(K)$ must be compact, so f must be proper. ∎

Another way to express the same idea, but that works even for unbounded domains, is to consider the one-point compactifications of the domains. Suppose that $f : \Omega_1 \to \Omega_2$ is a continuous mapping between arbitrary domains. One extends f to a mapping f^* between the compactifications by setting

$$f^* = f \text{ on } \Omega_1$$
$$f^*(\infty) = \infty. \tag{44}$$

The reader can verify that f is a proper mapping if and only if f^* is a continuous mapping between the compactifications.

In this book we will usually restrict our consideration to proper holomorphic mappings between bounded domains. The standard notation $\Omega \subset\subset \mathbb{C}^n$ denotes that Ω is a bounded domain in $\mathbb{C}^n$. The several reasons for this restriction will become clear as we proceed. In order to prove the central result of this section, we make a preliminary definition. Suppose that Ω is an open domain in some $\mathbb{C}^q$ and that

$$g : \Omega \to \mathbb{C}^n \tag{45}$$

is a nonconstant holomorphic mapping. We call either $g(\Omega)$ or g a parameterized complex analytic set. Later we will worry about the dimension of $g(\Omega)$, but for now this does not matter. When $q = 1$, we use the term "parameterized holomorphic curve," as the image is necessarily one-dimensional. To emphasize that g is not a constant, we sometimes say "nontrivial" parameterized complex analytic sets.

We are now ready to show that there is no proper holomorphic mapping from a polydisc to a ball. We will prove the following stronger statement.

THEOREM 11
Suppose that

$$\Omega_1 \subset\subset \mathbb{C}^{n_1}$$

$$\Omega_2 \subset\subset \mathbb{C}^{n_2}$$

$$\Omega \subset\subset \mathbb{C}^n = \mathbb{C}^{n_1 + n_2} \tag{46}$$

are all bounded domains. If there is a proper holomorphic mapping

$$f : \Omega_1 \times \Omega_2 \to \Omega \tag{47}$$

then $b\Omega$ *must contain nontrivial parameterized complex analytic sets.*

PROOF   We write (z, w) for the variables in $\mathbb{C}^n$. Suppose that f is the given proper holomorphic mapping. Choose an arbitrary point p in $b\Omega_1$ and let $\{z_k\}$ be a sequence tending to p. The sequence of $\mathbb{C}^n$-valued holomorphic mappings on Ω_2, defined by

$$w \to f(z_k, w), \tag{48}$$

is a bounded sequence in $O(\Omega_2, \Omega)$, because Ω is a bounded set. By Theorem 6, therefore, it has a convergent subsequence

$$w \to f(z_{k_m}, w) \tag{49}$$

whose limit function we denote by g_p. Since f is proper, the limit g_p must map, by Lemma 1, into $b\Omega$. If g_p is not constant, then $g_p(\Omega_2)$ is a parameterized analytic set in $b\Omega$. To finish the proof, we will show that such functions cannot all be constant.

Consider the mapping $(\partial f / \partial w)(z, w)$. As above, let z tend to an arbitrary boundary point p of Ω_1. By the convergence of $f(z_{k_m}, w)$ and the Cauchy estimates, the subsequence of matrix-valued functions on Ω_2 defined by the matrix of first derivatives

$$\frac{\partial f}{\partial w}(z_{k_m}, w) \tag{50}$$

also converges to $\partial g_p / \partial w$. If each g_p were a constant, then for each boundary point p, the corresponding sequence of derivatives would converge to 0. Now fix w. Since the boundary point is arbitrary,

$$\lim_{z \to b\Omega_1} \frac{\partial f}{\partial w} (z, w) = 0. \tag{51}$$

Thus any component function of (the matrix) $z \to (\partial f/\partial w)(z, w)$ is continuously extendable to $b\Omega_1$ with vanishing boundary values. By the maximum principle, it would be the zero function on Ω_1. This is absurd, because the vanishing of the derivative means that $f(z, w)$ would be independent of w. The inverse image of a point would then fail to be compact, and the mapping could not be proper. Thus there is some p for which the limit function g_p is not constant. This defines the nontrivial parameterized complex analytic set $g_p(\Omega_2)$ that lies in $b\Omega$. ∎

COROLLARY 3
There is no proper mapping from a polydisc to a ball.

PROOF By Theorem 11 it is enough to verify that there are no nontrivial parameterized complex analytic sets in the sphere. To do so, let U be an open, connected set in $\mathbb{C}$ and let $g : U \to S^{2n-1} \subset \mathbb{C}^n$. The function

$$t \to \|g(t)\|^2 \tag{52}$$

is then a constant. Applying the operator $(\partial/\partial t)(\partial/\partial \bar{t})$, we obtain that $\partial g/\partial t = 0$. Thus g is a constant, so every parameterized holomorphic curve in the sphere is trivial, and therefore so is every such complex analytic set. ∎

REMARK 1 In Section 7.1.3 we give a simple example of a biholomorphic mapping from an unbounded domain whose boundary contains parameterized holomorphic curves to a ball. This shows that the boundedness hypothesis in Theorem 11 is necessary. ∎

1.2.3 Solving analytic equations: Inverse and implicit functions

At this point in the preliminary chapter we turn to the holomorphic versions of the inverse and implicit function theorems. As mentioned in the preface, we will encounter situations where the hypotheses of these theorems are not necessarily satisfied. To understand such situations we first recall the fundamental theorems we must generalize.

It is well known in complex geometry that the real Jacobian determinant of a holomorphic mapping is the absolute value squared of its complex Jacobian determinant. We begin with a simple proof of this assertion. Its appearance here occurs because the hypotheses of invertibility of the derivative mappings in the real and complex cases are actually the same. It arises in complex geometry also in proving that a complex manifold has a natural orientation. In particular, a holomorphic mapping must be orientation preserving, a fact we need in Chapter 2. This fact also arises in Chapter 7 when we change variables in certain integrals.

LEMMA 2
Suppose that

$$f : \Omega \subset \mathbb{C}^n \to \mathbb{C}^n \tag{53}$$

is a holomorphic mapping. Let

$$F : \Omega \subset \mathbf{R}^{2n} \to \mathbf{R}^{2n} \tag{54}$$

denote the same mapping in the underlying real variables. Then the Jacobian determinants satisfy the relation

$$\det(dF) = |\det(df)|^2. \tag{55}$$

PROOF   Let z denote the domain variable and w the range variable. The volume forms in the domain and range are

$$dV(z) = c \, dz \wedge d\bar{z} = c \, dz^1 \wedge \cdots \wedge dz^n \wedge d\overline{z^1} \wedge \cdots \wedge d\overline{z^n}$$
$$dV(w) = c \, dw \wedge d\overline{w} \tag{56}$$

for the same constant c. Replacing dw^j by $\sum(\partial f_j/dz^k)dz^k$, doing the same for the conjugates, and using the definition of the determinant gives

$$dV(w) = |\det(df)|^2 \, dV(z). \tag{57}$$

Thus the volume distortion ratio is

$$|\det(df)|^2 \tag{58}$$

so this must also be the underlying real Jacobian determinant.   ∎

As a consequence of this lemma, a holomorphic mapping has an invertible derivative precisely when the underlying real mapping does. This enables us to derive the holomorphic versions of the inverse and implicit function theorems almost immediately from their real analogues. We state the real versions for continuously differentiable functions. Let E, E_1, E_2 denote real Banach spaces, which for us are finite-dimensional Euclidean spaces.

THEOREM 12
(Inverse function theorem, smooth version). *Suppose that*

$$F : \Omega \subset E \to E \tag{59}$$

is continuously differentiable and that, for some $p \in \Omega$, the derivative $dF(p)$ is invertible. Then there is a neighborhood of p on which F is one-to-one. The inverse function is also continuously differentiable, so the mapping is a local diffeomorphism.

THEOREM 13
(Implicit function theorem, smooth version). *Suppose that*

$$F : U \subset E_1 \times E_2 \to E_2 \tag{60}$$

is continuously differentiable and the linear mapping

$$\frac{\partial F}{\partial y}(x_o, y_o) : E_2 \to E_2 \tag{61}$$

is invertible. Here $(x, y) \in E_1 \times E_2$. Then there is a neighborhood of (x_o, y_o) and a continuously differentiable function G defined in a neighborhood of x_o such that $F(x, y) = F(x_o, y_o) \Leftrightarrow y = G(x)$ on these neighborhoods.

The proofs of these theorems appear in, for example, [La]. It is easy to derive the corresponding theorems in the holomorphic category from these theorems. At the risk of some redundancy we state the results separately, then easily reduce their proofs to Theorems 12 and 13.

THEOREM 14
(Inverse function theorem, holomorphic version). *Suppose that*

$$f : \Omega \subset \mathbb{C}^n \to \mathbb{C}^n \tag{62}$$

is a holomorphic mapping and that for some $p \in \Omega$, the derivative matrix $(\partial f/\partial z)(p)$ is invertible. Then there is a neighborhood of p on which f is one-to-one. The inverse function is also holomorphic, so f is a local biholomorphism.

PROOF Let F denote the corresponding mapping in the underlying real variables. By Lemma 2, $|\det(df(p))|^2 = \det(dF(p))$, so the corresponding derivative matrices are simultaneously invertible. Because $dF(p)$ is invertible, the smooth version of the theorem guarantees that there is a continuously differentiable local inverse F^{-1}. Since this inverse mapping is continuously differentiable, to show that it corresponds to a holomorphic mapping it is sufficient to verify the Cauchy–Riemann equations. Let z denote the variable in the domain and w the variable in the range. It follows from the chain rule (in matrix notation) and the Cauchy–Riemann equations for f, that

$$0 = \frac{\partial w}{\partial \overline{w}} = \frac{\partial f}{\partial z}\frac{\partial z}{\partial \overline{w}} + \frac{\partial f}{\partial \overline{z}}\frac{\partial \overline{z}}{\partial \overline{w}} = \frac{\partial f}{\partial z}\frac{\partial F^{-1}}{\partial \overline{w}}. \tag{63}$$

Since $\partial f/\partial z$ is invertible, we see that $\partial F^{-1}/\partial \overline{w} = 0$, so the inverse mapping also satisfies the Cauchy–Riemann equations. Thus it is holomorphic, and the theorem is proved. ∎

THEOREM 15

(Implicit function theorem, holomorphic case). *Suppose that*

$$f : \Omega \subset \mathbb{C}^{k+n} \to \mathbb{C}^n \tag{64}$$

is a holomorphic mapping and the linear mapping

$$\frac{\partial f}{\partial w}(z_0, w_0) : \mathbb{C}^n \to \mathbb{C}^n \tag{65}$$

is invertible. Here $(z, w) \in \mathbb{C}^k \times \mathbb{C}^n$. Then there are neighborhoods $U \ni (z_0, w_0)$ and $A \ni z_0$, and a holomorphic mapping $g : A \to \mathbb{C}^n$, such that $f(z, w) = f(z_0, w_0) \Leftrightarrow w = g(z)$ for $(z, w) \in U$.

PROOF Again we let F denote the corresponding mapping in the underlying real variables, and again the hypotheses on the derivative matrices correspond. Thus there are neighborhoods, as in the statement of the theorem, on which we can solve the equation $F(z, w) = F(z_0, w_o)$ for $w = g(z, \bar{z})$. The function g is continuously differentiable. To show that it is holomorphic, we differentiate the defining relation

$$f(z, g(z, \bar{z})) = f(z_0, w_o). \tag{66}$$

The chain rule and the Cauchy–Riemann equations yield

$$0 = \frac{\partial f}{\partial z}\frac{\partial z}{\partial \bar{z}} + \frac{\partial f}{\partial w}\frac{\partial g}{\partial \bar{z}} = \frac{\partial f}{\partial w}\frac{\partial g}{\partial \bar{z}}. \tag{67}$$

The matrix $\partial f/\partial w$ is invertible, so we must have $\partial g/\partial \bar{z} = 0$. As before, we see that a continuously differentiable mapping satisfies the Cauchy–Riemann equations; hence it must be holomorphic. ∎

It is useful to conceive of the implicit function theorem more geometrically. Another way to state this result is that n independent complex analytic equations in $k + n$ variables define locally k-dimensional complex analytic submanifold. For the implicit function theorem, independence is determined by the first derivatives. We will encounter more general situations where the derivative fails to have maximal rank, but for which the equations still behave essentially independently.

Suppose that Ω is open in $\mathbb{C}^{k+n}$ and that $f : \Omega \subset \mathbb{C}^{k+n} \to \mathbb{C}^n$ is a holomorphic mapping. The inverse image $f^{-1}(c)$ of a point in $\mathbb{C}^n$ is called a complex analytic subvariety of Ω. We use the notation $\mathbf{V}(f)$, pronounced "the variety defined by f," to denote the level set $f^{-1}(0)$, regardless of the number of component functions. The precise definition of a complex analytic subvariety is as follows.

DEFINITION 3 *Let B be an open subset of $\mathbb{C}^n$. A complex analytic subvariety of B is a subset V such that each point $p \in V$ has a neighborhood on which V is defined by the simultaneous vanishing of a finite set of holomorphic functions. This means that for each p we can find an open subset $\Omega = \Omega_p$ and holomorphic functions $f_j \in O(\Omega)$ such that*

$$V \cap \Omega = \{z \in \Omega : f_1(z) = \cdots = f_k(z) = 0\}. \tag{68}$$

A complex analytic variety is a topological space locally biholomorphic to a complex analytic subvariety. No difficulties will arise from our subsequent use of the terms "variety" and "subvariety" interchangeably. There are several conceivable definitions of the dimension of such an object; fortunately these definitions agree (see [GR]). Definition 3 does not demand that a complex analytic subvariety be "pure dimensional." In case there are pieces of various dimensions, one takes the dimension to be the maximal possibility.

Recall that a complex manifold is a topological space locally biholomorphic to an open subset of $\mathbb{C}^n$. Complex submanifolds of $\mathbb{C}^n$ are given locally by the vanishing of systems of holomorphic equations whose derivative matrices have the appropriate rank. Most points of a (pure dimensional) complex analytic subvariety turn out to be "regular" points, where the subvariety is a complex manifold; therefore it is usual to define the dimension of the subvariety to be the (complex) dimension of the manifold of regular points. The remaining points are the "singular" points. Roughly speaking we will say that n "independent" complex analytic equations in $m+n$ variables define an m-dimensional complex analytic subvariety. Conversely, if the level set $f^{-1}(c)$ of a mapping $f : B \subset \mathbb{C}^{m+n} \to \mathbb{C}^n$ has dimension m, then we may consider the defining functions to be independent. Geometers who study level sets of smooth functions cannot be so cavalier. We will be more precise about dimension when we discuss the local parameterization theorem and also in Chapter 2 when we consider "regular sequences."

1.2.4 Solving analytic equations: The Weierstrass preparation theorem

The famous Weierstrass "Vorbereitungssatz" enables us to solve equations when the conditions of the inverse function theorem fail. Suppose that f is holomorphic in a neighborhood of the origin in $\mathbb{C}^{n+1}$, vanishes at the origin, but does not vanish identically. When the linear term in the Taylor expansion for f vanishes, we cannot apply the implicit function theorem to solve for any of the variables. In this situation we can use the preparation theorem. It is convenient to view holomorphic functions as power series; this theorem also holds in quite general formal power series rings [Bk].

We state and prove the preparation theorem after introducing some of the standard notation for power series rings. Suppose that p is a point in $\mathbb{C}^n$. Recall that the germ of a holomorphic function at p is an equivalence class of holomorphic functions, where two functions are equivalent if they agree on some neighborhood of p. One considers germs of other classes of functions as well. We denote by

$$O = {}_nO_p = O_p \tag{69}$$

the ring of germs of holomorphic functions at p in $\mathbb{C}^n$. By the existence and uniqueness of the Taylor expansion it follows that this ring is isomorphic to the ring of convergent power series, denoted by $\mathbb{C}\{z - p\}$. Henceforth we will not distinguish $\mathbb{C}\{z - p\}$ from O_p. We say that the order of vanishing of an element $f \in O_p$ is m if all the nonzero terms in the power series are of degree at least m and there is some nonzero term of degree m.

Without loss of generality one usually supposes that the base point is the origin and omits it from the notation. In general one omits the dimension from the notation as well, but in the present section the method of proof involves induction on the dimension, and thus we will require the dimension in our notation. For purposes of the preparation theorem, we let

$$O = {}_{n+1}O \tag{70}$$

denote the ring of convergent power series at 0 in the $n + 1$ complex variables

$$(z, w) \qquad z \in \mathbb{C}^n, w \in \mathbb{C}. \tag{71}$$

A unit in this ring is a power series with nonzero constant term. A Weierstrass polynomial is a monic polynomial in w whose coefficients are convergent power series in z that vanish at 0. We write, as is customary, $O[w]$ for the ring of polynomials in w with coefficients in O.

THEOREM 16
(Weierstrass preparation theorem). *Suppose f is holomorphic near $(0,0)$ in $\mathbb{C}^{n+1}$,*

$$f(0,0) = 0, \tag{72}$$

but $f(0, w)$ does not vanish identically. Let m denote the order of vanishing of $f(0, w)$. Then there are a neighborhood of $(0,0)$ and holomorphic functions $u(z, w)$, $p(z, w)$ defined there, such that u is a unit, p is a Weierstrass polynomial of degree m, and

$$f(z, w) = u(z, w)p(z, w). \tag{73}$$

Such a decomposition is unique.

PROOF The standard proof goes as follows. The holomorphic function of one variable $w \to f(0, w) = F(w)$ has a zero of order m at the origin. Choose δ so that this function has exactly m zeroes inside $|w| < \delta$ and none on the circle. The number of roots, counting multiplicities, is given by the integral formula

$$\#Z(F) = \frac{1}{2\pi i} \int_{|t|=\delta} \frac{F'(t)}{F(t)} \, dt$$

$$= \frac{1}{2\pi i} \int_{|t|=\delta} \frac{\frac{\partial f}{\partial w}(0, t)}{f(0, t)} \, dt. \tag{74}$$

Treating z as a parameter, we see from this formula that the number of roots is locally constant in z. Thus, for a fixed z sufficiently close to zero, we can find m roots (counting multiplicities) to the equation $f(z, w) = 0$ for $|w| < \delta$. Calling these roots $\phi_j(z)$, which are not necessarily holomorphic functions of z, we nevertheless form the product

$$p(z) = \prod_{j=1}^{m} \left(w - \phi_j(z) \right). \tag{75}$$

The claim is that p is actually holomorphic and is the desired Weierstrass polynomial. To see this, multiply out the product, and note that the elementary symmetric functions of the $\phi_j(z)$ occur as coefficients. As these are polynomials in the power sums

$$\sum_{j=1}^{m} \left(\phi_j(z) \right)^d, \tag{76}$$

it is enough to prove that these power sums are holomorphic. To do so, consider the residue integral

$$\frac{1}{2\pi i} \int_{|t|=\delta} \frac{\partial f}{\partial w}(z, t) \frac{t^d}{f(z, t)} \, dt. \tag{77}$$

This integral evidently equals

$$\sum_{j=1}^{m} \left(\phi_j(z) \right)^d \tag{78}$$

by the one-variable residue theorem, if δ is chosen as above. Since $f(0, w) \neq 0$ on the circle of integration, the same holds by continuity for $f(z, w)$ if z is close enough to the origin. Therefore the residue integral defines a holomorphic function of z. Since (77) is holomorphic, so is (76), and hence p is a polynomial with holomorphic coefficients. By construction these coefficients vanish at

$z = 0$, so p is a Weierstrass polynomial. Next we consider the Cauchy integral

$$u(z, w) = \frac{1}{2\pi i} \int\limits_{|t|=\delta} \frac{f(z, t)}{p(z, t)} \frac{dt}{t - w} . \tag{79}$$

With δ chosen as above, this integral defines a holomorphic function because f and p have the same zeroes inside the contour. To verify that u is a unit, evaluate it at the origin, obtaining

$$u(0, 0) = \frac{1}{2\pi i} \int\limits_{|t|=\delta} \frac{f(0, t)}{t^{m+1}} \, dt, \tag{80}$$

which does not vanish because $f(0, t)$ has a term of order m in its Taylor expansion. Finally, to see that $f = up$, multiply both sides of (79) by $p(z, w)$ and apply the Cauchy integral formula.

The uniqueness is easy to check, so its proof will be left to the reader. ∎

The preparation theorem generalizes a special case of the implicit function theorem. Suppose we have precisely one equation; in the notation of Theorem 15, $n = 1$. That theorem applies when there is a nonzero linear term in the Taylor expansion, i.e., $m = 1$. Its conclusion is that the equation $f(z, w) = 0$ becomes equivalent (locally) to the equation $p(z, w) = w - a(z) = 0$. The preparation theorem applies also when $m > 1$. We can find m local solutions $w = w(z)$ to the equation $f(z, w) = 0$, and these solutions are the roots of a (Weierstrass) polynomial equation. Although the solutions are not necessarily holomorphic, the "root system" is holomorphic. See [Wh] for an exposition from this point of view.

REMARK 2 If f is any holomorphic function vanishing at the origin, but not identically zero, then one can always make a linear change of coordinates so that the hypotheses of the preparation theorem are satisfied. See [GR] for this fact and generalizations. ∎

Suppose that f vanishes to order m at the origin. According to Remark 2, we can always make a linear coordinate change so that the hypotheses of the preparation theorem apply. It is also sometimes convenient to apply the theorem to a fixed variable, for which the order of vanishing may be larger. Suppose for example that we put $f(z, w) = z^3 + w^4 - zw$. Then f is a Weierstrass polynomial in z of order 3 and in w of order 4. In a generic linear direction, it vanishes to order 2. By Remark 2, it can also be written as a unit times a Weierstrass polynomial of degree 2 in appropriate coordinates.

1.2.5   The Weierstrass division theorem and consequences

The Weierstrass division theorem arises when one considers integrals of the form (79), but when f and p do not have the same zeroes. The result is as follows:

THEOREM 17

(Weierstrass division theorem). *Suppose that $f = f(z, w)$ is holomorphic near the origin in $\mathbb{C}^{n+1}$ and that p is a Weierstrass polynomial in w of degree m. Then there are unique holomorphic functions q, r such that the remainder r is a polynomial in w of degree less than m, and*

$$f = pq + r. \tag{81}$$

PROOF   Again we leave the uniqueness to the reader. Given f and p, we form the same integral as before:

$$q(z, w) = \frac{1}{2\pi i} \int\limits_{|t|=\delta} \frac{f(z,t)}{p(z,t)} \frac{dt}{t - w}. \tag{82}$$

This defines a holomorphic function if we make sure the contour of integration avoids the zero set of the Weierstrass polynomial p. Then we compute

$$
\begin{aligned}
f(z, w) - p(z, w)q(z, w) &= r(z, t) \\
&= \frac{1}{2\pi i} \int\limits_{|t|=\delta} \frac{f(z,t)p(z,t) - f(z,t)p(z,w)}{p(z,t)(t - w)} \, dt
\end{aligned}
\tag{83}
$$

where we have written f as a Cauchy integral and added fractions. Factoring the numerator shows that $(t - w)$ divides it. Since $p(z, w)$ is a polynomial in w, we see that

$$r(z, w) = \frac{1}{2\pi i} \int\limits_{|t|=\delta} \frac{f(z,t)}{p(z,t)} g(z, t - w) \, dt \tag{84}$$

where g is a polynomial in its second variable. Since this is the only dependence of $r(z, w)$ on w, it is a polynomial in w as well. ∎

There are many consequences of these two theorems of Weierstrass. For us the most important are the algebraic properties of the ring of convergent power series. One important obvious property of this ring is that it is a local ring; this means that the non-units form the unique maximal ideal M. Thus M is the ideal generated by the germs of the coordinate functions. Not so obvious are the Noetherian property and the unique factorization into irreducible elements. The word "ring" in this book means commutative ring with an identity element.

DEFINITION 4   *A ring is called local if it has a unique maximal proper ideal. A ring is called Noetherian if every ideal is finitely generated. A ring is called a unique factorization domain if every element admits a finite factorization into irreducible elements, and this factorization is unique up to units and the order of the factors.*

THEOREM 18
The local ring $O = O_p$ is Noetherian.

PROOF   We suppose that p is the origin. Let I be an ideal in O. The assertion is trivial if $I = (0)$ or if I is the whole ring. Otherwise find an element of I that is not identically zero but that vanishes at the origin. After a linear change of coordinates, and by the preparation theorem, we can assume that this element is a Weierstrass polynomial p of degree m in $w = z_n$. By the division theorem, each element f in I can be written $f = pq + r$, where r is a polynomial in $w = z_n$ of degree less than m. Each such remainder lies in both I and in $_{n-1}O[w]$. Let us denote by J the intersection of the ideal I with the polynomial ring:

$$J = I \cap {}_{n-1}O[w]. \tag{85}$$

It is easy to check that J is an ideal in the ring $_{n-1}O[w]$. This suggests an inductive proof, which goes as follows:

The ring $_1O$ is Noetherian as it is obviously in fact a principal ideal domain. Assume that we have proved that $_{n-1}O$ is a Noetherian ring. The Hilbert basis theorem then implies that the ring $_{n-1}O[w]$ is also Noetherian. Therefore, for any ideal

$$I \subset {}_nO, \tag{86}$$

the ideal J constructed above is finitely generated. Then I is generated by the Weierstrass polynomial p and the generators of J, and is thus itself finitely generated.   ∎

THEOREM 19
The ring $_nO$ is a unique factorization domain.

PROOF   The proof is again by induction on the dimension; the one-dimensional case is immediate. Again the inductive step is facilitated by the use of the polynomial ring $_{n-1}O[w]$. By the induction hypothesis and a lemma of Gauss, this ring is also a unique factorization domain. Now, given an element f in $_nO$, we may assume (after a linear change of coordinates) that it is a unit times a Weierstrass polynomial in w. Factorizing the Weierstrass polynomial into irreducible elements in $_{n-1}O\,[w]$, we can write

$$f = u \prod_{j=1}^{k} f_j. \tag{87}$$

This is also a factorization of f into irreducible elements in $_nO$. Suppose that we have another factorization

$$f = v \prod_{j=1}^{m} g_j. \tag{88}$$

By evaluating $f\,(0, w)$ we see that each $g_j\,(0, w) \neq 0$. Thus we can apply the preparation theorem to each of these, obtaining a factorization

$$f = v \prod_{j=1}^{m} g_j = v' \prod_{j=1}^{m} p_j. \tag{89}$$

The factorizations (87) and (88) are both factorizations in $_{n-1}O\,[w]$, so by the induction hypothesis they must agree up to units. Thus (87) gives the unique factorization of f into irreducibles. ∎

Thus the local ring $_nO$ is a Noetherian unique factorization domain. In Section 1.4 we discuss some deeper properties of this ring. We close this section by remarking that, although each ideal is finitely generated, the number of generators of an ideal in $_nO$ can be arbitrarily large for a fixed $n \geq 2$. For example, the ideal M^k that is generated by the collection of (germs of) homogeneous monomials of degree k requires

$$\binom{k + n - 1}{n - 1}$$

generators.

1.3 Further applications

1.3.1 Local subvarieties of codimension one

The Weierstrass preparation theorem enables us to study the local zero set of a holomorphic function by studying the vanishing of a Weierstrass polynomial. This has the advantage of permitting the use of elementary ideas from algebra. The purpose of the present section is to use some algebra to prove the intuitively obvious statement that "most" points of a codimension-one complex analytic subvariety are smooth points. This fact is necessary in the proof of the biholomorphicity of an injective equidimensional holomorphic mapping, which is the subject of the next section. We are concerned only with local properties, so we use the word "variety" instead of the more pedantic "germ of an analytic subvariety" for the local zero set of a germ of a holomorphic function.

The local ring $_{n-1}O$ is an integral domain, since the product of nonvanishing power series does not vanish. Therefore it has a field of fractions, denoted by $_{n-1}F$. The ring $_{n-1}F[w]$ is then a polynomial ring in one variable over a field, and thus it is easy to understand. In particular, if $f, g \in {}_{n-1}F[w]$, then we can apply the Euclidean algorithm from elementary arithmetic to find their greatest common divisor (f, g). The greatest common divisor is a single generator for the ideal generated by the pair. The following elementary lemma seems omnipresent in the consideration of the zeroes of polynomials.

LEMMA 3

An element g in $_{n-1}F[w]$ has no multiple roots if and only if

$$\left(g, \frac{\partial g}{\partial w} \right) = 1 \tag{90}$$

in $_{n-1}F[w]$.

PROOF   (Sketch). If w_o is a multiple root, then $w - w_o$ divides both g and $\partial g/\partial w$, so their greatest common divisor cannot be unity. Conversely, the Euclidean algorithm enables us to explicitly compute the greatest common divisor. The process decreases the degrees of the polynomials until a nonzero element of $_{n-1}F$ is reached, as long as there are no multiple roots. Dividing through, we obtain the result.   ∎

Let us investigate the zero locus of a local holomorphic function. By the preparation theorem it is enough to do this for a Weierstrass polynomial p. By unique factorization we can factor the polynomial into irreducible factors,

$$p = \prod_{j=1}^{k} p_j, \tag{91}$$

each of which is also a Weierstrass polynomial. From this it is evident that the zero locus $\mathbf{V}(p)$ can be written

$$\mathbf{V}(p) = \bigcup_{j=1}^{k} \mathbf{V}(p_j). \tag{92}$$

Thus it suffices to study the zero locus of an irreducible Weierstrass polynomial.

LEMMA 4

Let p be an irreducible Weierstrass polynomial in $_{n-1}O[w]$. Then there are *elements A, B, q in $_{n-1}O[w]$ such that*

$$Ap + B\frac{\partial p}{\partial w} = q \tag{93}$$

and such that q does not vanish identically on the (germ of a) variety $\mathbf{V}(p)$.

PROOF Since p is irreducible, it has no multiple roots. By Lemma 3, therefore, there is an identity of the form

$$A_1 p + A_2 \frac{\partial p}{\partial w} = 1 \tag{94}$$

for $A_1, A_2 \in {}_{n-1}F[w]$. The least common denominator q of the coefficients of A_1, A_2 is a nonzero element of $_{n-1}O$. Multiplying through by this common denominator yields an identity of the form

$$Ap + B\frac{\partial p}{\partial w} = q. \tag{95}$$

Since q is independent of w and does not vanish identically, we can find a point z_o where $q(z_o) \neq 0$. Since $w \to p(z_o, w)$ is a nonconstant polynomial, it does have zeroes. Thus q does not vanish everywhere that p does. ∎

PROPOSITION 2
Let p be an irreducible Weierstrass polynomial, and let q be as in Lemma 4. Let

$$\pi : \mathbb{C}^{n-1} \times \mathbb{C} \to \mathbb{C}^{n-1}$$

$$\pi(z, w) = z \tag{96}$$

be the indicated projection. Then the (dense) subset

$$\mathbf{V}(p) - \pi^{-1}(\mathbf{V}(q)) \subset \mathbf{V}(p) \tag{97}$$

consists of regular points. That is, $\mathbf{V}(p)$ is a complex submanifold of codimension one near any point in this set.

PROOF Suppose that

$$(z, w) \in \mathbf{V}(p) - \pi^{-1}(\mathbf{V}(q)). \tag{98}$$

Then

$$p(z, w) = 0$$

$$q(z, w) = q(z) \neq 0 \Rightarrow \frac{\partial p}{\partial w}(z, w) \neq 0 \tag{99}$$

by formula (93) for q. By the (holomorphic) implicit function theorem, the point (z, w) is a regular point of

$$\mathbf{V}(p) - \pi^{-1}(\mathbf{V}(q)) \tag{100}$$

and the result is proved. $\blacksquare$

DEFINITION 5   *A subset $K \subset \mathbb{C}^n$ is called thin, if, for each point $p \in K$, there is a neighborhood $U \ni p$ and a nonconstant holomorphic function $h : U \to \mathbb{C}$ such that $K \cap U \subset \mathbf{V}(h)$.*

Thus a thin subset is locally contained in the zero set of a holomorphic function. It is easy to prove the statement that the complement in a domain Ω of the zero set of a holomorphic function is connected. Such a zero set is therefore too small to disconnect a domain; this gives geometric meaning to the word "thin." After one defines holomorphic function on a variety, one makes an analogous definition of "thin" subset of a variety. We now show that the singular points of a codimension-one subvariety also form a "thin" subset of the variety.

THEOREM 20
If V is any codimension-one (germ of a) complex analytic subvariety in $\mathbb{C}^n$, then the set of singular points of V is contained in the zero set of a nonzero holomorphic function on V and is thus a "thin" set.

PROOF   We may suppose by the preparation theorem that the variety is defined by a Weierstrass polynomial. Factor this into finitely many irreducible Weierstrass polynomials $p = \prod p_i^{m_i}$. We may suppose that no factors are repeated because the variety defined is unchanged if all the m_i equal unity. Thus the variety is a finite union of "branches" $\mathbf{V}(p) = \cup \mathbf{V}(p_i)$. On each branch $\mathbf{V}(p_i)$, Proposition 2 guarantees the existence of an appropriate function q_i. Then $q = \prod q_i$ does not vanish identically on the variety, and the complement on V of its zero set consists only of smooth points. $\blacksquare$

Example 1
We can follow the proof for specific polynomials. Put

$$p(z, w) = w^2 - z_1 z_2. \tag{101}$$

The corresponding function q can be chosen to be

$$q(z) = z_1 z_2 \tag{102}$$

because, as ideals,

$$\left(p, \frac{\partial p}{\partial w}\right) = \left(w^2 - z_1 z_2, 2w\right) = (w, z_1 z_2). \tag{103}$$

Thus we simplify the generators of the ideal by inspection to obtain an element independent of w. One can also follow the algorithm of the preceding proof, and write

$$\frac{1}{-z_1 z_2}\left(w^2 - z_1 z_2\right) + \frac{w}{2z_1 z_2}(2w) = 1 \tag{104}$$

to see that the least common denominator of the coefficients is the function

$$q(z) = z_1 z_2. \tag{105}$$

It is interesting to observe that the pair of lines $(t, 0, 0)$ and $(0, t, 0)$ lie in the zero sets of both p and $\partial p / \partial w$, even though the singular locus consists of the origin alone. This does not contradict Theorem 20; it simply means that the singular locus can be even thinner than the thin set found by the proof. □

In Section 4 we generalize these results to subvarieties of higher codimension. Before doing so we give a nice application of the fact that most points on a codimension-one subvariety are regular points.

1.3.2 An injective equidimensional holomorphic mapping is biholomorphic

In any advanced calculus course that mentions the inverse function theorem, it is necessary to mention also the polynomial function

$$f(x) = x^3 \tag{106}$$

that is one-to-one on the real line, but whose inverse is not even differentiable. It is remarkable that such a difficulty cannot occur in the holomorphic category. It is an old result in several complex variables that an injective holomorphic equidimensional mapping is necessarily biholomorphic. The proof has evolved to a fairly simple induction argument [GH,Ro1].

THEOREM 21
Suppose that $f : \Omega \subset \mathbb{C}^n \to \mathbb{C}^n$ is holomorphic and one-to-one. Then f is a biholomorphism onto its image.

PROOF By hypothesis f is globally one-to-one. It is therefore sufficient to show that f is locally biholomorphic to obtain the conclusion of the theorem. To do so, it is sufficient to verify that $df(p)$ is invertible for each point p, for the inverse function theorem then applies.

We prove this by induction on the dimension. The one-variable case is easy. If we wish to solve $f(z) = w$ near the point p, then we write

$$f(z) - f(p) = (z - p)^m u(z) = w - f(p) \tag{107}$$

where $u(p)$ does not vanish and m is a positive integer. The integer m completely describes the local behavior. The function is locally m-to-one; hence the hypothesis guarantees that m is unity, and therefore $df(p) = u(p) \neq 0$.

Suppose now that we have proved the theorem for all dimensions up to and including $n - 1$. We need an auxiliary result. Fix $p \in \Omega$. The needed result is that the rank of $df(p)$ must be 0 or n. To prove this, assuming the induction hypothesis, suppose that the rank is positive. Choose notation so that $\partial f_n(p) \neq 0$, and then make a linear change of coordinates so that

$$\frac{\partial f_n}{\partial z_k}(p) = 0, \ k < n \quad \text{but} \quad \frac{\partial f_n}{\partial z_n}(p) \neq 0. \tag{108}$$

The set

$$\{z \in \Omega : f_n(z) = f_n(p)\} \tag{109}$$

is then a complex submanifold M (of dimension $n - 1$) of a neighborhood of p by the implicit function theorem. We may identify a neighborhood of p within it as an open subset of $\mathbb{C}^{n-1}$. The restriction g of f to this neighborhood remains one-to-one, so by the induction hypothesis of the theorem, $dg(p)$ has rank $n - 1$. (Note that the last component of f must be omitted to ensure that g is an equidimensional mapping.) But

$$df(p) = \begin{pmatrix} dg(p) & * \\ 0 & \frac{\partial f_n}{\partial z_n} \end{pmatrix} \tag{110}$$

by construction, and hence has full rank by the induction hypothesis and (108). This shows that the auxiliary result holds. To repeat, either $df(p)$ is invertible, or it vanishes.

We finish the proof by verifying that $df(p)$ cannot have rank zero. Consider the branch locus defined by

$$Z = \mathbf{V}(\det(df)). \tag{111}$$

If Z is not empty, then it is a codimension-one complex analytic subvariety. By Theorem 20 it must contain a regular point p. Let us consider a neighborhood in Z of this regular point, on which all points are regular, and examine the derivative of f. By hypothesis, the derivative is not invertible there, so by the auxiliary result,

$$df(z) = 0 \tag{112}$$

for all points in this neighborhood. But then the restriction of f to an open subset of Z is constant, which contradicts the assumption that f is one-to-one. Hence Z is empty and the theorem is proved. ∎

The restriction to the same dimension is essential in this theorem. The variety that arises in the following example appears several times in this book; it serves often as the simplest counterexample to a naive conjecture.

Example 2
Put

$$z(t) = \left(t^2, t^3\right). \tag{113}$$

Then z is a homeomorphism from $\mathbb{C}$ to the one-dimensional subvariety of $\mathbb{C}^2$ defined by $V = \{z_1^3 - z_2^2 = 0\}$. This mapping is one-to-one, yet its derivative vanishes at the singular point $t = 0$. More generally, by adding variables to this example, we see that the auxiliary result also fails for mappings that are not equidimensional. $\quad\square$

1.4   Basic analytic geometry

1.4.1   Ideals and germs of varieties

The purpose of this section is to discuss the interplay between complex analytic varieties and ideals. Associated with the germ of a complex analytic variety is an ideal, and associated with an ideal in O is the germ of an analytic subvariety. These constructions enable us to pass between geometry and algebra. One important result is the (Rückert) Nullstellensatz; this generalizes to holomorphic germs the (Hilbert) Nullstellensatz for polynomials. The Nullstellensatz is fundamental to the rest of this book, for the following reason. We will reduce questions about the contact of real and complex varieties to questions about ideals of holomorphic functions. The Nullstellensatz determines when such ideals define trivial varieties. Before proceeding to the requisite algebra, we recall some language concerning germs of complex analytic varieties.

The language of germs enables us to study the geometry of a variety near a given point. Suppose V_j is a complex analytic subvariety of a neighborhood Ω_j of the point p, for $j = 1, 2$. We say that V_1 is equivalent to V_2 at p, if there is a neighborhood Ω of p on which $V_1 \cap \Omega = V_2 \cap \Omega$. Equivalence of varieties at a point is an equivalence relation, yielding the notion of the germ (V, p).

DEFINITION 6   *A germ at p of a complex analytic subvariety is an equivalence class of complex analytic subvarieties under the equivalence relation just described.*

Often the point in question will be the origin. Thus $(V, 0)$ denotes the germ of V at 0, although we will not be systematic in distinguishing germs from their representatives. The notation $f : (V, 0) \rightarrow (V', 0)$ means that f is (the germ of) a holomorphic mapping satisfying $f(0) = 0$ and such that $f(V) \subset V'$.

We now begin to discuss the relationship between geometry and algebra. Recall that O denotes the (Noetherian) ring of germs of holomorphic functions at the origin. Associated with germs of subvarieties are ideals in O and conversely. One important idea is that different ideals may define the same zero set, and we must take this into account. This motivates the notion of the radical of an ideal.

Given that I is an ideal in O, we associate with it the germ of a variety $\mathbf{V}(I)$ defined by $\{z : f(z) = 0 \; \forall f \in I\}$. More precisely, we proceed as follows. Since I is finitely generated, we can find representatives $f_1^*, \ldots, f_k^*$ of the germs of these generators that are defined on a common open set Ω. Then $\mathbf{V}(I)$ is the germ represented by the subvariety defined by $\{z \in \Omega : f_1^*(z) = \ldots = f_k^*(z) = 0\}$. Next, given that V is the germ of a variety, we associate with it the ideal $\mathbf{I}(V)$ defined by $\mathbf{I}(V) = \{f \in O : f(z) = 0 \; \forall z \in V\}$. More precisely, a germ g lies in $\mathbf{I}(V)$ if, on some open set, there is a holomorphic function g^* representing g and an analytic subvariety V^* representing V, such that $g^*(z) = 0 \; \forall z \in V^*$.

It is evident that each of these operations reverses set theoretic containment. It is also clear that $\mathbf{V}(\mathbf{I}(V)) = V$ always holds. In general $\mathbf{I}(\mathbf{V}(I)) \supset I$, but equality need not hold. It is a deep theorem, the Rückert Nullstellensatz, that $\mathbf{I}(\mathbf{V}(I)) = \mathrm{rad}(I)$, where the radical is defined in Definition 7. Theorem 22 states the Nullstellensatz in elementary but equivalent language.

DEFINITION 7   *If I is an ideal in O, then its radical is the ideal defined by*

$$\mathrm{rad}(I) = \{g : \exists k : g^k \in I\}. \tag{114}$$

One easily verifies the following formal facts about varieties, ideals, and radicals:

$$\mathrm{rad}(I \cap J) = \mathrm{rad}(I) \cap \mathrm{rad}(J)$$

$$\mathbf{V}(I \cap J) = \mathbf{V}(I) \cup \mathbf{V}(J)$$

$$\mathbf{I}(V_1 \cup V_2) = \mathbf{I}(V_1) \cap \mathbf{I}(V_2)$$

$$\mathbf{V}(I) = \mathbf{V}(\mathrm{rad}(I)). \tag{115}$$

The importance of the radical derives from the Nullstellensatz. This theorem says that if g vanishes on the set of common zeroes of elements in I, then it lies in the radical of I. We repeatedly use the Nullstellensatz in the special case when the set of common zeroes of I consists of one point. Such ideals play a fundamental role. Recall that an ideal I is called M-primary, or primary to the maximal ideal, if it satisfies the containment

$$\mathbf{M}^k \subset I \subset \mathbf{M} \tag{116}$$

for some k. By the Nullstellensatz, this is equivalent to the statement that I defines a trivial variety, since it implies that $\text{rad}(I) = M$.

THEOREM 22

(Nullstellensatz). *Let $f_1, \ldots, f_k$ be holomorphic functions, defined near the origin and vanishing at the origin. Suppose g is a holomorphic function that vanishes on the set $\mathbf{V}(f)$. Then there is an integer N so that g^N lies in the ideal $(f) = (f_1, \ldots, f_k)$. Equivalently, for any ideal $I \subset O$, $\mathbf{I}(\mathbf{V}(I)) = \text{rad}(I)$.*

We give the proof for germs of analytic functions in Section 4.3. The Nullstellensatz holds, and is much easier to prove, for complex polynomials. In Chapter 6 we state an analogue for real analytic, real-valued functions, due to Lojasiewicz. There the notion of radical requires inequalities rather than equalities for its definition. There is even a generalization to formal power series rings.

The proof of the Nullstellensatz breaks into two steps. First one makes multiple use of the Weierstrass preparation and division theorems to prove the theorem for prime ideals. The general case then follows from purely algebraic considerations. We continue with these algebraic notions.

DEFINITION 8   *An ideal I is called prime if*

$$fg \equiv 0 \bmod (I), \quad g \not\equiv 0 \bmod (I) \Rightarrow f \equiv 0 \bmod (I). \tag{117}$$

It is called primary if

$$fg \equiv 0 \bmod (I), \quad g \not\equiv 0 \bmod (I) \Rightarrow \exists N : f^N \equiv 0 \bmod (I), \tag{118}$$

that is, $f \in \text{rad}(I)$.

An example of a primary ideal that is not prime is the ideal (z^m) (for $m > 1$) in one variable. The importance of primary ideals is the primary decomposition theorem [V] to the effect that every ideal in a Noetherian ring is a finite intersection of primary ideals. We show next that the Nullstellensatz for prime ideals implies the general case.

LEMMA 5

Given that $\mathbf{I}(\mathbf{V}(Q)) = Q$ holds for all prime ideals Q in O, it follows that $\mathbf{I}(\mathbf{V}(I)) = \text{rad}(I)$ holds for all ideals I in O.

PROOF   Note that the radical of a primary ideal must be itself a prime ideal. Suppose we knew that, for prime ideals Q, we must have $\mathbf{I}(\mathbf{V}(Q)) = Q$. Then, for any primary ideal P, we have $\mathbf{I}(\mathbf{V}(P)) = \mathbf{I}(\mathbf{V}(\text{rad}(P))) = \text{rad}(P)$, where we have used (115) and the statement for prime ideals. Now suppose that I is any ideal. By the primary decomposition theorem, we can write it as a finite intersection of primary ideals. First doing so, and then applying (115), we obtain

$$I = \cap P_a \Rightarrow \mathbf{V}(I) = \mathbf{V}(\cap P_a) = \cup \mathbf{V}(P_a)$$

$$\Rightarrow \mathbf{I}(\mathbf{V}(I)) = \mathbf{I}(\cup \mathbf{V}(P_a)) = \cap \mathbf{I}(\mathbf{V}(P_a)). \qquad (119)$$

Assuming that $\mathbf{I}(\mathbf{V}(P_a)) = \mathrm{rad}(P_a)$, we obtain

$$\mathbf{I}(\mathbf{V}(I)) = \cap \mathrm{rad}(P_a) = \mathrm{rad}(\cap P_a) = \mathrm{rad}(I), \qquad (120)$$

which is the desired conclusion. ∎

Thus the Nullstellensatz for an arbitrary ideal follows from the Nullstellensatz for a primary ideal, which in turn follows from the (simpler) result for prime ideals. To prove the theorem for prime ideals requires the theory of coherent analytic sheaves, some version of the local parameterization theorem, or other deep work. See Section 4.3 for a proof that relies on our sketch of the local parameterization theorem, or see [Gu1] or [GR] for complete proofs.

Before turning to the local parameterization theorem, we consider briefly the notion of reducibility for a variety. This gives further evidence of the importance of prime ideals. We saw already that a complex analytic subvariety of codimension one is the union of branches defined by irreducible Weierstrass polynomials. There is a similar statement for germs of complex analytic subvarieties of any codimension.

DEFINITION 9 *A complex analytic variety V is called reducible if it can be written as a nontrivial union*

$$V = V_1 \cup V_2 \qquad (121)$$

of complex analytic subvarieties, and irreducible otherwise. We have the same definition for germs.

By nontrivial union we mean that neither variety in (121) is empty. It is a simple consequence of the Noetherian property of O that every germ of a complex analytic subvariety can be uniquely written as a finite irredundant union of irreducible germs of complex analytic subvarieties. To see this, suppose that V is a given germ; if it is irreducible, then the result holds. If it is reducible, then we can write $V = V_1 \cup V_2$. If each of these germs is irreducible, then the result holds. Otherwise, at least one of these varieties is itself reducible. Continue in this manner. Suppose the process does not terminate. Then we have a strictly decreasing sequence of varieties under containment. Taking ideals, we obtain a strictly increasing sequence of ideals in O. This is impossible in a Noetherian ring. Therefore the process terminates, and we obtain a decomposition into finitely many irreducible branches. The reader can easily verify that this decomposition is unique up to the order of the branches.

It is worth remarking that there is a distinction between irreducibility of a variety and irreducibility of its germ at a point. It is easy to verify that the germ $(V, 0)$ is irreducible precisely when $\mathbf{I}(V)$ is a prime ideal. See [Gu1] for additional information.

1.4.2 Statement of the local parameterization theorem

In this section we state the local parameterization theorem for complex analytic varieties. This theorem makes precise the sense in which the germ of a complex analytic subvariety is a "branched covering" over a complex Euclidean space of lower dimension. Several times in the sequel we use some of the conclusions of this theorem. Although we refer to [Gu1] for a complete proof, we give some explicit examples that illustrate the situation quite nicely. The reader can develop his intuition by first considering the codimension-one case, and recognizing that the following theorem is but an attempt to generalize that intuition to the case of higher codimension. See [Gu1,GR] for more details.

THEOREM 23
Suppose that V is an irreducible germ at the origin of a complex analytic subvariety in $\mathbb{C}^n$. Then there is a system of coordinates, an integer q, a neighborhood B of the origin in $\mathbb{C}^n$, and a thin subset $E \subset \pi(B) \subset \mathbb{C}^q$ such that all the following hold.

1. *Let*

$$\pi(z_1, \ldots, z_n) = (z_1, \ldots, z_q) \tag{122}$$

    be the projection (in the coordinates above). Then

$$V \cap B - \pi^{-1}(E) \tag{123}$$

    is a q-dimensional complex analytic submanifold that is dense in V.

2.

$$\pi : V \cap B - \pi^{-1}(E) \twoheadrightarrow \pi(B) - E \tag{124}$$

    is a finitely sheeted covering mapping.

3. *The set*

$$V \cap B - \pi^{-1}(E) \tag{125}$$

    is connected.

4. *The topological dimension of V as a set at the origin equals q.*

In Proposition 2, concerning an irreducible germ of codimension one, we obtained the thin set E as the zero set of the least common divisor of a Weierstrass polynomial and its derivative. An analogous statement holds in general. To see this, we make several uses of the preparation theorem. We assume that $\mathbf{I}(V)$ is prime.

Let P be a proper prime ideal. Since P is not the zero ideal, there is a minimal order of vanishing m_n of elements of P. Since P is not the whole ring, $m_n \geq 1$. After a linear change of coordinates, we can find an element $f_n \in P$ that is regular of order m_n in z_n. By the preparation theorem, we see that there is a Weierstrass polynomial $p_n \in {}_{n-1}O\,[z_n] \cap P$ of this degree. Either ${}_{n-1}O \cap P = 0$ or the process continues. If it continues, then we select an element of minimal order m_{n-1} in ${}_{n-1}O \cap P$. Again we obtain a Weierstrass polynomial. Continue this process. We see that there are coordinates $(z_1, \ldots, z_n)$, an integer $q \geq 0$, and Weierstrass polynomials $p_j \in {}_{j-1}O\,[z_j]$ such that

$$ {}_qO \cap P = 0 $$

$$ p_k \in {}_{k-1}O\,[z_k] \cap P \quad \text{for} \quad k = q+1, \ldots, n. \tag{126} $$

Here ${}_kO$ denotes the ring of germs of holomorphic functions in the first k variables. Only linear coordinate changes are involved. For prime ideals, the integer q is independent of the choice of coordinate system.

Note that

$$ (p_{q+1}, \ldots, p_n) \subset P \tag{127} $$

so that

$$ \mathbf{V}(p_{q+1}, \ldots, p_n) \supset \mathbf{V}(P). \tag{128} $$

The projection (122) maps $\mathbf{V}(P)$ properly to $\mathbb{C}^q$.

To get more information we must use the condition that P is prime. Following the ideas in Section 3.1 we let ${}_qF$ denote the quotient field of the integral domain ${}_qO$. Consider also the quotient field F of the integral domain ${}_nO/P$. The existence of the Weierstrass polynomial p_n reveals that the residue class ζ_n of z_n in ${}_nO/P$ satisfies a monic polynomial equation, that is, ζ_n is integral. Proceeding with the variables in decreasing order, we see that the quotient ring ${}_nO/P$ is an integral algebraic extension of ${}_qO$. Therefore F is an algebraic extension of ${}_qF$, obtained by adjoining the residue classes of each z_j for $j \geq q+1$. Using some elementary field theory, namely the theorem of the primitive element [V], we can realize this extension by adjoining one element, which is a linear combination of the residue classes of the z_j. By making one more linear coordinate change, one can therefore guarantee the following. The residue class of z_{q+1} generates the field extension from ${}_qF$ to F. Assuming this has been done, we consider the particular Weierstrass polynomial p_{q+1} in z_{q+1}. We compute its discriminant $d \in {}_qO$ defined by

$$ d = \left(p_{q+1}, \frac{\partial p_{q+1}}{\partial z_{q+1}} \right). \tag{129} $$

Thus d is the greatest common divisor of p_{q+1} and its derivative. We emphasize that d depends only on $(z_1, \ldots, z_q)$ and is not in the ideal P. As before, the thin set E in Theorem 23 is the zero set of d. We give more information in the proof of the Nullstellensatz.

In the circumstances of Theorem 23, we say that $\pi : \mathbf{V}(P) \to \mathbb{C}^q$ is a branched analytic covering. Away from the zero locus of d, the variety $\mathbf{V}(P)$ is a complex manifold of dimension q. The first q variables serve as local coordinates and parameterize the variety there. The sheets of the variety can coalesce in rather complicated ways along the branch locus.

REMARK 3 We have seen already in Example 1 that the subset $\pi^{-1}(E)$ may be larger than the set of singular points. The complement of $\pi^{-1}(E)$ must be open and dense. $\blacksquare$

We consider two additional examples of explicit varieties where the origin is a singular point.

Example 3
Put

$$V = \mathbf{V}\left(z_3^2 + 2z_1 z_3 + 2z_1 z_2\right). \tag{130}$$

There are two branches of the "function" π^{-1}; these are defined by

$$\pi_+^{-1}(z_1, z_2) = \left(z_1, z_2, -z_1 + \left(z_1^2 - 2z_1 z_2\right)^{1/2}\right)$$

$$\pi_-^{-1}(z_1, z_2) = \left(z_1, z_2, -z_1 - \left(z_1^2 - 2z_1 z_2\right)^{1/2}\right). \tag{131}$$

The thin set $E \subset \mathbb{C}^2$ is given by

$$E = \mathbf{V}\left(z_1^2 - 2z_1 z_2\right) = \mathbf{V}(z_1) \cup \mathbf{V}(z_1 - 2z_2). \tag{132}$$

Again the inverse image $\pi^{-1}(E)$ is larger than the singular locus of V, which consists of the origin alone. One can also check easily that the holomorphic mapping ϕ defined by

$$(s, t) \to \phi(s, t) = \left(\frac{-s}{2(1+t)}, st, s\right) \tag{133}$$

maps into V, but is not locally surjective, as the lines given by

$$(0, t, 0) \quad \text{and} \quad (t, 0, 0) \tag{134}$$

lie in V but are not in the image of the mapping. $\square$

Example 4

Define a subvariety V of $\mathbb{C}^4$ by the equations

$$z_1 z_4^2 - \left(\frac{z_3}{\sqrt{5}}\right)^5 = 0$$

$$z_1^3 z_4 - \left(\frac{z_2}{\sqrt{5}}\right)^5 = 0. \tag{135}$$

This variety will arise in our study of proper mappings between balls in Chapter 5. The situation is rather complicated because the ideal is not prime. One checks for instance that there are divisors of zero of the form $z_1^2 z_4 - c_j z_3 z_2^3$ for five distinct constants c_j. These functions are not in the ideal, but their product is. Next let us compute the singular locus of V. The matrix of derivatives is

$$\begin{pmatrix} z_4^2 & 0 & cz_3^4 & 2z_1 z_4 \\ 3z_1^2 z_4 & cz_2^4 & 0 & z_1^3 \end{pmatrix}. \tag{136}$$

Any point of V for which $z_2 z_3 \neq 0$ is a smooth point by the complex implicit function theorem, because the middle two-by-two matrix is then invertible. If either of these variables vanishes at a point of the variety, then the other must also. If both vanish, then we see that V contains the one-dimensional complex lines given by

$$(\zeta, 0, 0, 0) \quad \zeta \in \mathbb{C}$$

$$(0, 0, 0, \zeta) \quad \zeta \in \mathbb{C}. \tag{137}$$

These are obviously smooth, and the rank of the derivative matrix is unity at all points along these branches except at the origin. At the origin these two lines join up with the smooth two-dimensional manifolds defined by

$$\left(\frac{t^2}{s}, \sqrt{5}\,\epsilon t, \sqrt{5}\,\delta s, \frac{s^3}{t}\right) \quad s, t \in \mathbb{C}^* \quad \epsilon^5 = \delta^5 = 1. \tag{138}$$

Actually the variety is purely two-dimensional. The parameterization (138) cannot be specialized to give these lines, but we can set, for example, $\zeta = t^2/s$, eliminate s, and obtain another rational parameterization of the form

$$\left(\zeta, \sqrt{5}\,\epsilon t, \sqrt{5}\,\delta \frac{t^2}{\zeta}, \frac{t^5}{\zeta^3}\right) \quad \zeta, t \in \mathbb{C}^* \quad \epsilon^5 = \delta^5 = 1. \tag{139}$$

Setting $t = 0$ in (139) now gives the first complex line. We obtain the second in a similar manner. Returning to (138), we note that if $t \sim s^\alpha$, $1/2 < \alpha < 3$, then z_1, z_4 become continuous functions of s, t at the origin. If we define $\pi : \mathbb{C}^4 \to \mathbb{C}^2$ by $\pi(z_1, z_2, z_3, z_4) = (z_2, z_3)$, then none of the determinations of π^{-1} is continuous at the origin. $\quad\square$

REMARK 4   Suppose that V is as in the statement of the local parameterization theorem. Suppose also that h is holomorphic in a neighborhood of the origin, and that

$$h = 0 \quad \text{on} \quad \pi_j^{-1}(\pi(B) - E) \tag{140}$$

where π_j^{-1} is one of the inverses of π. Then h vanishes on V.   ∎

1.4.3   The Nullstellensatz

We discover more information about the discriminant in the following proof of the Nullstellensatz. This author learned a similar proof from a course taught by Gunning. See [Gu1] for detailed information.

PROOF OF THEOREM 22   According to Lemma 5, it is sufficient to prove that $\mathbf{I}(\mathbf{V}(P)) = P$ when P is prime. It is sufficient to show that $\mathbf{I}(\mathbf{V}(P)) \subset P$, as the opposite inclusion is trivial. We will accomplish this as follows. We will write an arbitrary element f of O as

$$d^N f \equiv t_f \bmod P \tag{141}$$

for an appropriate integer N, discriminant d, and polynomial $t_f \in {}_qO\left[z_{q+1}\right]$. When f is in $\mathbf{I}(\mathbf{V}(P))$, this polynomial will be forced to vanish identically. It then follows that

$$d^N f \equiv 0 \bmod P \tag{142}$$

and therefore, since P is prime, that $f \equiv 0 \bmod P$. Thus we begin by proving (141).

   Choose coordinates and the integer q as in (126), so that we have in particular the Weierstrass polynomial p_{q+1}. From it we compute the discriminant function d defined by (129). This does not lie in the ideal P, since ${}_qO \cap P = 0$. Let us write m_j for the degree of the polynomials p_j.

   Let f be an arbitrary element of O. By the division theorem, $f = p_n s_n + r_n$, so $f \equiv r_n \bmod P$, where the remainder is of degree less than m_n. Consider a coefficient of this remainder. We also apply the division theorem to it, using the Weierstrass polynomial p_{n-1}. Again we may replace it by a polynomial r_{n-1} of degree less than m_{n-1}. We continue this process. Each time a variable z_j for $j > q+1$ occurs as a coefficient, we can similarly replace it by a polynomial of degree less than m_j. The result is that we can write

$$f \equiv w \bmod P \tag{143}$$

where $w \in {}_qO\left[z_{q+1}, \ldots, z_n\right]$ is a polynomial whose degree in each variable is less than the corresponding m_j.

Next we want to eliminate completely the variables z_j for $j > q + 1$. To do so, we claim that there are elements $h_k \in {}_qO\left[z_{q+1}\right]$ so that

$$z_k d \equiv h_k\left(z_{q+1}\right) \bmod P. \tag{144}$$

The proof of (144) amounts to solving a system of linear equations. The determinant of the system (145) is the square of the discriminant. For a fixed point $z' \in \mathbb{C}^q$, we look at the inverse images $\pi_j^{-1}\left(z'\right)$, $j = 1, \ldots, m_q$. Denote the kth coordinate of one of these inverses by $\left(\pi_j^{-1}\left(z'\right)\right)_k$. To simplify notation, let $m = m_q$. Consider the system of linear equations

$$d\left(z'\right)\left(\pi_s^{-1}\left(z'\right)\right)_k = \sum_{j=0}^{m-1} a_{jk}\left(z'\right)\left(p_{q+1}\left(\pi_s^{-1}\left(z'\right)\right)\right)^j \tag{145}$$

for the matrix of unknowns $a_{jk}\left(z'\right)$. There is an equation for each of the inverses π_s^{-1}, $s = 1, \ldots, m$. Using matrix notation we write (145) as

$$\left(F_{sk}\right) = \left(G_{sj}\right)\left(a_{jk}\right). \tag{146}$$

We can solve (145) precisely when

$$\det\left(\left(p_{q+1}\left(\pi_s^{-1}\right)\left(z'\right)\right)^j\right) = \det\left(G_{sj}\right) \neq 0. \tag{147}$$

Since the matrix $\left(G_{sj}\right)$ is a Vandermonde matrix, its determinant is the square of the discriminant. By allowing the factor d on the left, we are guaranteed the invertibility as long as there are points where d does not vanish. Put $a_{jk}\left(z'\right) = \left(G^{js}\right)\left(F_{sk}\right)$ and then set

$$h_k\left(z', t\right) = \sum_{j=0}^{r-1} a_{jk}\left(z'\right) t^j. \tag{148}$$

We substitute (148) in (143) and eliminate all variables z_k for $k > q + 1$. For an appropriate integer N, we obtain (141), where t_f is a polynomial in z_{q+1} of degree lower than m_{q+1}. Its coefficients are in ${}_qO$.

Now we finally suppose also that $f \in \mathbf{I}\left(\mathbf{V}\left(P\right)\right)$. From (141) so is t_f. Away from the zero set of d, however, $\mathbf{V}(P)$ has m_{q+1} sheets. For a fixed $z' = \left(z_1, \ldots, z_q\right)$ in this good set, $t_f\left(z', z_{q+1}\right)$ vanishes m_{q+1} times. The polynomial t_f then vanishes more times than its degree, so it is forced to be identically zero. Putting this in (141) yields (142). As we remarked at the beginning of the proof, since P is prime, we obtain that $f \in P$, so $\mathbf{I}\left(\mathbf{V}\left(P\right)\right) \subset P$. ∎

REMARK 5   We can choose (instead of p_{q+1}) any function $g \in {}_qO\left[z_{q+1}\right]$ whose values are distinct, to ensure that the discriminant does not vanish. The condition that the residue class of z_{q+1} generate the field extension is equivalent to this assertion. ∎

Example 5

In case the integer q equals zero, we see that there is a power of z_1 in the ideal. Since the ideal is prime, in fact z_1 itself is in the ideal. Working backwards, we see that the ideal must have been the maximal ideal. Thus the maximal ideal is the only prime ideal that defines a trivial variety. $\Box$

To determine whether an ideal is prime can be tricky, although Theorem 2.3 gives a simple method when the ideal defines a one-dimensional variety.

Example 6

Consider the ideal $I = \left(z_3^2 - z_1^3, z_2^2 - z_1^5 \right)$. It is not prime. One way to see this is to notice that

$$(z_2 - z_1 z_3)(z_2 + z_1 z_3) \in I \tag{149}$$

but that the factors are not in I. Another way is to notice that the variety

$$\mathbf{V}\left(z_3^2 - z_1^3, z_2^2 - z_1^5 \right) \tag{150}$$

consists of two complex analytic curves through the origin, which is an isolated singular point. These curves are parameterized by $\left(t^2, \pm t^5, t^3 \right)$.

By way of contrast, consider the ideal $J = \left(z_3^2 - z_1^5, z_2^3 - z_1^7 \right)$. This ideal is prime. It defines the parameterized curve $\left(t^6, t^{14}, t^{15} \right)$. The discriminant is a power of z_1, and the origin is an isolated singular point. $\Box$

2

Holomorphic Mappings and Local Algebra

2.1 Finite analytic mappings

2.1.1 Introduction

One aim of this book is to study the geometry of a real hypersurface in complex Euclidean space. The main idea will be to assign to each point on the hypersurface a family of ideals of holomorphic functions. The information contained in these ideals is sufficient to decide, for example, whether the hypersurface contains complex analytic varieties, and, if so, of what dimension. By computing numerical invariants of these ideals, we will obtain quantitative information on the order of contact of complex analytic varieties with the hypersurface. To understand this construction, it is natural to begin with a study of the geometry of ideals of holomorphic functions. This vast subject deserves a more detailed treatment than is given here. We present (somewhat informally) the material required in Chapters 3 and 4; we also include other information that reveals the depth and unity of the subject.

This chapter aims particularly toward understanding the properties of those ideals of holomorphic functions that define isolated points. Suppose therefore that we are given a collection of holomorphic functions $(f) = (f_1, \ldots, f_N)$ vanishing at the origin. Consider the system of equations $f(z) = 0$. How do we decide whether this system admits nontrivial solutions? A (germ of a) finite analytic mapping (see Definition 1) is a mapping for which the origin is an isolated solution to this system. We consider also quantitative aspects which we now motivate.

The theory is considerably simpler when the range and domain are of the same dimension. Suppose therefore that we are given a holomorphic mapping f defined near the origin in $\mathbb{C}^n$ and taking values also in $\mathbb{C}^n$. When the origin is an isolated solution to the equations $f(z) = 0$, we wish to measure an analogue of the "order of the zero" there. This amounts to the determination of "how many times" the given equations define the origin. Any injective mapping defines the origin only once. For other equidimensional finite analytic mappings, the analogue of "order of the zero" will be the generic number of inverse images of nearby points. Theorem 1 displays many ways to compute this number. We require other measurements in case the domain and range are of different dimensions.

We return to the general case. Suppose that the system of equations $f(z) = 0$ has nontrivial solutions. We want first to determine the dimension of the (germ of a) variety $\mathbf{V}(f)$ defined by these equations. Suppose that the dimension q of $\mathbf{V}(f)$ is positive. Recall that we use the word "branch" to mean irreducible variety. Several types of singularities for $\mathbf{V}(f)$ are possible. There may be more than one branch passing through the origin. It may also happen that each branch of $\mathbf{V}(f)$ can be defined by simpler equations; we then seek to measure the "number of times" that the given equations $f(z) = 0$ define each branch. A branch of $\mathbf{V}(f)$ may be singular, even though its defining equations are as simple as possible. Again we want to give a quantitative measurement of the singularity.

We measure these considerations by assigning various numbers to ideals in O. Let us describe some properties these numbers should satisfy. Fix the dimension n of the domain. For each integer k satisfying $1 \leq k \leq n$, we wish to assign a number (or $+\infty$), written $\#_k(f)$, to the germ of (f) at the origin. This measurement should be finite precisely when $\dim(\mathbf{V}(f)) < k$; it should equal unity precisely when $\operatorname{rank}(df(0)) > n - k$, and it should increase as the singularity becomes more pronounced. In case $n = N = 1$, the number $\#_1(f)$ should equal the order of the zero. In general the numbers should take into account both how many branches the variety has and also how many times each branch is defined. Finally, we should have the inequalities

$$\#_n(f) \leq \#_{n-1}(f) \leq \ldots \#_1(f). \tag{1}$$

There are many possible choices; we will study the geometric meaning of some of these choices and the relationships among them.

We now turn to finite analytic mappings; for these mappings $\#_1(f) < \infty$. The term mapping is intended both to convey geometric intuition and to emphasize that the range of the function will in general be of dimension larger than one. By treating mappings rather than complex-valued functions as the basic object, one can recover many of the basic properties of holomorphic functions of one variable. The simplest example concerns solution sets. A nontrivial function of one variable has isolated zeroes. Suppose that k holomorphic functions of n variables vanish at a common point. The dimension of the analytic subvariety defined by them must be at least $n - k$. It is therefore necessary to consider mappings

$$f : \Omega \subset \mathbb{C}^n \to \mathbb{C}^N \tag{2}$$

where $N \geq n$, in order that they have isolated zeroes. Such mappings are of vital importance in the rest of this book.

DEFINITION 1 *Suppose that*

$$f : \Omega \subset \mathbb{C}^n \to \mathbb{C}^N \qquad (3)$$

is a holomorphic mapping. Then f is called a finite analytic mapping if, for each $w \in f(\Omega)$, the fiber $f^{-1}(w)$ is a finite set.

Sometimes we use the terms "finite holomorphic mapping" or "finite mapping" with the same meaning as "finite analytic mapping." Note that proper holomorphic mappings between domains in complex Euclidean spaces are finite mappings. This follows because the inverse image of a point is necessarily a compact complex analytic subvariety, and the only such subvarieties of domains in $\mathbb{C}^n$ are finite sets [Ra,Ru1].

Recall that the notation $f : (\mathbb{C}^n, p) \to (\mathbb{C}^n, 0)$ denotes that f is the germ of a holomorphic mapping from $\mathbb{C}^n$ to $\mathbb{C}^N$ and that $f(p) = 0$. A germ $f : (\mathbb{C}^n, p) \to (\mathbb{C}^N, 0)$ defines a finite analytic mapping when $f^{-1}(0) = p$ as "germs of sets." This means that there is a neighborhood of p on which p is the unique solution to the equation $f(z) = 0$. In some situations an N-tuple of holomorphic functions defined near p will be considered as a holomorphic mapping whose range is a subset of $\mathbb{C}^N$. Sometimes this N-tuple should be construed as a representative for the germ of this mapping. In other situations, an N-tuple will represent generators for the ideal that it generates in ${}_n O_p$. The context will make clear which of these interpretations is appropriate.

Our immediate concerns are to determine when a given mapping is finite and to measure the multiplicity of the solution. For both questions it suffices to study the ideal generated by the components. It is standard in algebraic and analytic geometry to consider intersection numbers in this context. A common thread in this book will be the reduction of questions about the contact of real and complex varieties to questions about intersection numbers for ideals of holomorphic functions. In Chapter 3 we recall the definition of the Levi form on a real hypersurface. When this form degenerates, there are many possible ways to measure the singularity. Algebraic methods enable us to understand the relationships among these measurements.

From the point of view of commutative algebra, the nicest situation occurs when one has the same number of equations as variables, so we will study this case in some detail. Suppose that $f : (\mathbb{C}^n, 0) \to (\mathbb{C}^n, 0)$ is a finite analytic mapping; in this case the (germ of the) variety $V(f)$ consists of just one point. According to our stated aims, all the numbers $\#_q(f)$ should then be finite. To motivate their definitions we recall some elementary facts from the one-dimensional case.

2.1.2 The one-dimensional case

Because the power series ring in one variable is a principal ideal domain, the one-dimensional case is trivial but illuminating. Let f be a holomorphic function defined near 0 in $\mathbb{C}$. There are precisely three possibilities. First, f can vanish identically, a case of little interest. Here we assign the number infinity: $\#(f) = \infty$. Second, f can be a unit in the ring of germs of holomorphic functions at 0; in other words, $f(0) \neq 0$. In this case, also of little interest, we put $\#(f) = 0$. The third alternative is that $f(0) = 0$ and that the zero is of finite order m. We put $\#(f) = m$. It is worthwhile to consider the many interpretations of the order of vanishing.

The algebraic point of view considers the function as the generator of an ideal. As ideals,

$$(f) = (z^m) \tag{4}$$

for the same exponent m as above. The quotient space $O/(f)$ is then a finite-dimensional algebra. Thought of as a vector space, it has a basis consisting of

$$\left\{ 1, z, z^2, \ldots, z^{m-1} \right\} \tag{5}$$

and is thus m-dimensional.

The geometric point of view considers the mapping properties of the function. For a point w sufficiently close to 0, there are m preimages z that satisfy $f(z) = w$. One way to see this is to write

$$f(z) = z^m u(z) \tag{6}$$

where u is a unit. Since $u(0) \neq 0$, we can define a holomorphic mth root of u close to 0. Taking mth roots, and since

$$h(z) = z(u(z))^{\frac{1}{m}} \tag{7}$$

is holomorphic and locally one-to-one, we find the m solutions $h^{-1}(w^{\frac{1}{m}})$ to equation (6).

The topological approach considers winding numbers and integral formulas. Again write

$$f(z) = z^m u(z) \tag{8}$$

and suppose that δ is small enough that $u(z) \neq 0$ for $|z| \leq \delta$. Then the number m is the value of a line integral:

$$\frac{1}{2\pi i} \int\limits_{|z|=\delta} \frac{f'(z)}{f(z)} \, dz = \frac{1}{2\pi i} \int\limits_{|z|=\delta} \frac{m}{z} \, dz + \frac{1}{2\pi i} \int\limits_{|z|=\delta} \frac{u'(z)}{u(z)} \, dz$$

$$= m + 0 = m. \tag{9}$$

Recall more generally that a closed contour γ is a (continuous) piecewise smooth mapping $\gamma : [a, b] \to \mathbb{C}$ such that $\gamma(a) = \gamma(b)$. If such a closed contour does not pass through the origin, then its winding number about the origin is

$$N(\gamma) = \frac{1}{2\pi i} \int\limits_{\gamma} \frac{d\zeta}{\zeta} . \tag{10}$$

Suppose that $\gamma = f\eta$ is the image under f of a positively oriented simple closed curve η about the origin. By changing variables and using the residue theorem, we obtain

$$\frac{1}{2\pi i} \int\limits_{\gamma} \frac{d\zeta}{\zeta} = \frac{1}{2\pi i} \int\limits_{\eta} \frac{f'(z)}{f(z)} \, dz = m. \tag{11}$$

Thus the number m also represents the winding number of the image curve γ about the origin. Note that we used this formula in the proof of the Weierstrass preparation theorem.

Closely related to these ideas is Rouché's theorem. This standard result tells us the following. If $|g| < |f|$ on a simple closed curve η, then we can consider g to be a small perturbation of f and conclude that the number of zeroes of $f + g$ inside η equals that of f inside η. In particular, if the origin is a zero of order m for f, then we may put $g(z) = -w$ for some small w and see that the equation $f(z) = w$ has m solutions. This technique works also for equidimensional finite analytic mappings.

In one dimension we thus see that all these approaches yield the same integer, and the last two show us that a nonconstant holomorphic function is an open mapping. At the risk of excessive repetition, we define the measurement # by the equation $\#(f) = m$. It follows that $\#(g) < \infty$ holds precisely when g does not vanish identically, that is, when $\dim(\mathbf{V}(g)) < 1$. Furthermore, $\#(g) = 1$ precisely when $\mathbf{V}(g)$ consists of the origin, defined just once.

2.1.3 The index of a mapping

To begin our general discussion of equidimensional finite analytic mappings, we first consider the notion of winding numbers in several variables. See [AGV] for more information. It is convenient to begin with smooth mappings on real Euclidean space. Suppose Ω is an open subset of real Euclidean space $\mathbf{R}^m$, and $F : \Omega \to \mathbf{R}^m$ is a smooth mapping. Suppose also that $a \in \Omega$ is an isolated point of the set $F^{-1}(0)$. Choose a sphere Ξ centered at a small enough so that no other roots lie inside or on Ξ. The mapping

$$\frac{F}{\|F\|} : \Xi \to S^{m-1} \tag{12}$$

is then defined and continuous. From the point of view of homology, we identify Ξ with the unit sphere. Recall that the $(m-1)$st homology group of S^{m-1} is the integers:

$$H_{m-1}\left(S^{m-1}\right) \simeq \mathbf{Z}. \tag{13}$$

The continuous mapping (12) then induces a mapping on the homology group **Z**; this induced mapping is a group homomorphism and therefore must be multiplication by some integer. This integer is commonly called the topological degree of the mapping (12) or of the mapping F. Its value is independent of the choice of sphere Ξ.

DEFINITION 2 *Let $F : (\mathbf{R}^m, a) \to (\mathbf{R}^m, 0)$ be the germ of a smooth mapping, and suppose that a is an isolated point in $F^{-1}(0)$. The index of F at the point a, written $\mathrm{index}_a(F)$, is the topological degree of (12).*

The simplest example is an orthogonal linear transformation T. Such a transformation preserves the sphere, and either preserves or reverses orientation. Hence $\deg(T) = \det(T)$. Thus $\mathrm{index}_0(T)$ equals the determinant and hence can be negative. For a holomorphic mapping, however, the index cannot be negative. This fact follows from Lemma 1.2, which states essentially that a holomorphic mapping must be orientation preserving, and lies behind many of the ideas in this chapter.

A version of the change of variables formula for multiple integrals holds for mappings with finite topological degree. We state without proof a simple version of that formula:

$$\mathrm{index}_0(F) \int\limits_{||x|| \le \epsilon} dV = \int\limits_{||F(x)|| \le \epsilon} \det(dF)\, dV. \tag{14}$$

Here both integrals have the same orientation. This formula holds for mappings that are generically d-to-one, where $d = \mathrm{index}_0(F)$, when the set of points with fewer than d inverse images has measure zero. Note also from (14) that $\det(dF)$ cannot vanish identically.

COROLLARY 1
The index of the germ of a holomorphic mapping $f : (\mathbb{C}^n, 0) \to (\mathbb{C}^n, 0)$ is positive.

PROOF Let $F : (\mathbf{R}^{2n}, 0) \to (\mathbf{R}^{2n}, 0)$ be the underlying real mapping. From Lemma 1.2, we have $\det(dF) = |\det(df)|^2$. From (14) we see therefore that $\mathrm{index}_0(f) = \mathrm{index}_0(F) > 0$. ∎

It is clear from the definition of index and from the discussion in one dimension that there is much to be gained from consideration of deformations or perturbations of a given mapping. In particular we can spread apart zeroes that have coalesced. It is necessary to recall the notion of continuous and smooth perturbations of mappings. Suppose that $0 \in \Omega \subset \mathbf{R}^k$ and that $F : \Omega \to \mathbf{R}^m$ is a smooth mapping. Suppose that Y is a parameter space that contains a base point; for most purposes the parameter space can be treated as a real interval containing the origin. We consider mappings $\phi : \Omega \times Y \to \mathbf{R}^m$ and define our perturbations F_ϵ by $F_\epsilon(x) = \phi(x, \epsilon)$. We demand that ϕ be continuous or smooth, and that $\phi(x, 0) = F(x)$.

In case $f = \sum c_\alpha z^\alpha$ is the germ of a holomorphic mapping, we usually think of $f_\epsilon(z) = \sum c_\alpha(\epsilon) z^\alpha$; that is, we allow the Taylor coefficients to depend smoothly on the parameter. As in Rouché's theorem from basic complex analysis, we also obtain perturbations as follows. Suppose that $f : (\mathbb{C}^n, 0) \to (\mathbb{C}^n, 0)$ and $g : (\mathbb{C}^n, 0) \to (\mathbb{C}^n, g(0))$ are germs of holomorphic mappings satisfying $\|g\| < \|f\|$ on some sphere about the origin. We consider the homotopy $f + \lambda g$, for $\lambda \in [0, 1]$, between f and $f + g$. All the mappings in this family will then have the same index about the origin.

For any smooth (or continuous) deformation f_ϵ, the mappings $f_\epsilon / \|f_\epsilon\|$ are all homotopic. This enables us to compute the index of f. Suppose $f(a) = 0$ but $f(z) \neq 0$ for $0 < \|z - a\| \leq \delta$. For sufficiently small ϵ, f_ϵ will be a smooth deformation of f that has finitely many zeroes in $\{z : \|z - a\| \leq \delta\}$. Denote these zeroes by a_i. The degree of $f_\epsilon / \|f_\epsilon\|$ then equals both the degree of $f / \|f\|$ and also the sum of the indices at these zeroes. This yields the coalescing property for the index of f.

PROPOSITION 1

The index $\mathrm{index}_a(f)$ *of a holomorphic mapping at an isolated root equals the sum of the indices of the roots of a deformation:*

$$\sum_{\{f_\epsilon = 0\}} \mathrm{index}_{a_i}(f_\epsilon) = \mathrm{index}_a(f). \tag{15}$$

We observe in Proposition 4 that the index for holomorphic mappings measures the generic number of preimages of points sufficiently close to the origin. If a is a point in the domain of f, and a is an isolated point in $f^{-1}(0)$, then the index is defined. Consider also the case where f has finitely many roots a_j inside a fixed ball B. Each root contributes to the topological degree of the mapping $f / \|f\| : bB \to S^{2n-1}$ between spheres, as described by the next proposition.

PROPOSITION 2

Suppose $B \subset \mathbb{C}^n$ is a ball, and $f : B \to \mathbb{C}^n$ is holomorphic, continuous on bB, and does not vanish there. If the mapping $f / \|f\| : bB \to S^{2n-1}$ has degree k, then f has at most k roots in B and

$$k = \sum_{f(a_i) = 0} \mathrm{index}_{a_i}(f). \tag{16}$$

PROOF   First we show that f has fewer than $k + 1$ roots. Suppose that f has roots at a_j for $j = 1, \ldots, k + 1$. We can find a polynomial mapping P with

$P\left(a_j\right) = 0$ $\forall j$ and such that $f + \epsilon P$ has nondegenerate zeroes for a generic small ϵ. The mapping

$$\frac{f + \epsilon P}{\|f + \epsilon P\|} : bB \to S^{2n-1}$$

also has degree k, as it is homotopic to f. Put a small ball B_j about each root a_j. The degree about each is $+1$ because the roots are nondegenerate. Consider the boundary Ξ of the domain $B \backslash \cup B_j$. Each sphere bB_j is negatively oriented with respect to this domain. Next consider the degree of the mapping

$$\frac{f + \epsilon P}{\|f + \epsilon P\|} : \Xi \to S^{2n-1}.$$

There is a contribution to the degree of $+k$ from bB. There is a contribution of -1 from each bB_j, yielding a contribution of $-(k+1)$. There may be other negative contributions if the small balls contain other roots. In any case, other contributions cannot be positive. Hence

$$\frac{f + \epsilon P}{\|f + \epsilon P\|} : \Xi \to S^{2n-1}$$

has degree at most -1. This contradicts Corollary 1, because $f + \epsilon P$ is holomorphic. This proves that the number of roots of f is at most k.

To obtain (16) we again consider $f_\epsilon = f + \epsilon P$. We assume that each root a_j of f is a nondegenerate root of f_ϵ, but we allow f_ϵ to have other roots. Since the mappings are homotopic, the degrees in (17) are the same. Putting this together with the coalescing property (15) of the index, applied to each root of f, we see that

$$
\begin{aligned}
k = \deg\left(\frac{f}{\|f\|}\right) &= \deg\left(\frac{f + \epsilon P}{\|f + \epsilon P\|}\right) \\
&= \sum_{\{f_\epsilon = 0\}} \text{index}_{b_j}\,(f + \epsilon P) \\
&= \sum_{\{f = 0\}} \text{index}_{a_j}\,(f). \qquad (17)
\end{aligned}
$$

This completes the proof. ∎

The argument in the proof of Proposition 2 is easy to visualize for one complex variable, and it appears, for example, in proofs of the residue theorem or the argument principle. In the next sections we will see how to calculate the index both by integration and by the techniques of commutative algebra. In particular, the index or winding number of a mapping will depend only on the ideal generated by its component functions in the ring of germs of holomorphic functions.

We continue to develop the analogy between equidimensional finite mappings and nonconstant functions in one dimension. Proposition 3 demonstrates one important similarity. The simple example $f(z) = (z_1, z_1 z_2)$ reveals that equidimensional mappings are not in general open mappings.

PROPOSITION 3

An equidimensional finite analytic mapping is an open mapping.

PROOF   Suppose without loss of generality that $f : (\mathbb{C}^n, 0) \to (\mathbb{C}^n, 0)$ is a finite map germ. Since $\|f\|^2 \geq c > 0$ on a sphere Ξ about zero, the degree of $f/\|f\| : \Xi \to S^{2n-1}$ is defined. It equals some positive integer k. To verify the openness of the mapping, we need to show that all points sufficiently close to the origin are in the range of f. Before doing so we indicate the role of k. According to Proposition 2, no point has more than k preimages, and according to Proposition 4 below most points near the origin have precisely k preimages. Fix the small sphere Ξ about the origin, on which $\|f\|$ is bounded away from zero. For every w satisfying $\|w\| < \inf_\Xi \|f(z)\|$, the mappings $f/\|f\|$ and $(f - w)/\|f - w\|$ have the same degree. This number is positive for $f/\|f\|$ and hence must also be positive for $(f - w)/\|f - w\|$. This can hold only when the denominator vanishes somewhere, so there must be some z for which $f(z) = w$. Hence every point w near the origin is in the range of f and the conclusion holds.   ∎

For a smooth map germ $F : (\mathbf{R}^m, 0) \to (\mathbf{R}^m, 0)$ on real Euclidean space, a point y in the range is called a "regular value" if $y = F(x)$ for some x satisfying $\det(dF)(x) \neq 0$. A point in the range is otherwise called a "critical value," as it is the image of a critical point. By Sard's theorem [La], the set of critical values has measure zero. The generic point in the range is therefore a regular value. Suppose that f is a holomorphic map germ, for which F is the underlying real mapping. We combine Sard's theorem with the coalescing property to obtain our next interpretation of the index.

PROPOSITION 4

The index $\text{index}_a(f)$ *of the germ of a holomorphic mapping* $f : (\mathbb{C}^n, a) \to (\mathbb{C}^n, 0)$ *is equal to the generic number of preimages* $\#(f^{-1}(w))$ *of values* w *sufficiently close to the origin.*

PROOF   Consider again the perturbations $f - w$. It follows from the coalescing property (15) that $\text{index}_a(f) = \sum_{f^{-1}(w)} \text{index}_z(f - w)$. By Proposition 3, every w is a value. If w is a regular value, then each term in the sum equals $+1$, because $f - w$ is holomorphic. Therefore $\text{index}_a(f) = \#(f^{-1}(w))$ if w is a regular value. By Sard's theorem this holds for a generic w.   ∎

2.1.4 Dolbeault cohomology and the Bochner–Martinelli integral

In this section we first recall some elementary facts about Dolbeault cohomology. This will suggest an integral formula for the index of a finite analytic mapping. Let us write $\Lambda^{p,q}(\Omega)$ for the smooth differential forms of type (p,q) on a domain $\Omega \subset \mathbb{C}^n$. We drop the domain from the notation when it is understood. The $\bar{\partial}$-closed $(p,0)$ forms are denoted by $O^{p,0}$. The Cauchy–Riemann operator $\bar{\partial}$ defines a complex

$$0 \to O^{p,0} \to \Lambda^{p,0} \to \Lambda^{p,1} \to \cdots \to \Lambda^{p,n} \to 0. \qquad (18)$$

The Dolbeault cohomology groups are defined as usual by the $\bar{\partial}$-closed forms modulo the $\bar{\partial}$-exact forms:

$$H^{p,q} = Z^{p,q}/B^{p,q}. \qquad (19)$$

Here

$$Z^{p,q} = \left\{ u \in \Lambda^{p,q} : \bar{\partial}u = 0 \right\}$$
$$B^{p,q} = \left\{ \omega \in \Lambda^{p,q} : \omega = \bar{\partial}g \, , \, g \in \Lambda^{p,q-1} \right\}. \qquad (20)$$

Perhaps the most important result in the theory of functions of several complex variables is the statement that

$$H^{p,q}(\Omega) = 0 \; \forall q \geq 1 \qquad (21)$$

if and only if Ω is a pseudoconvex domain. It is considerably easier to verify that these groups vanish for polydiscs. See [Hm,Ra].

To understand winding numbers and residues in several variables, it helps to know that, for $n \geq 2$,

$$\dim \left(H^{p,n-1} \left(\mathbb{C}^n - \{0\} \right) \right) = \infty, \quad 0 \leq p \leq n. \qquad (22)$$

Let us verify this explicitly when $p = n = 2$. The technique arises also in our computations with the Bochner–Martinelli integral.

Example 1
Write

$$\mathbb{C}^2 - \{0\} = U_1 \cup U_2$$

$$U_1 = \{(z,w) : z \neq 0\}$$
$$U_2 = \{(z,w) : w \neq 0\}. \qquad (23)$$

For each pair of nonnegative integers, we consider the meromorphic function $1/z^a w^b$. Then, in the overlap $U_1 \cap U_2$,

$$
\begin{aligned}
0 = \bar{\partial}\left(\frac{1}{z^a w^b}\right) &= \bar{\partial}\left(\frac{|z|^{2a} + |w|^{2b}}{\left(z^a w^b\right)\left(|z|^{2a} + |w|^{2b}\right)}\right) \\
&= \bar{\partial}\left(\frac{\bar{z}^a}{w^b\left(|z|^{2a} + |w|^{2b}\right)}\right) + \bar{\partial}\left(\frac{\bar{w}^b}{z^a\left(|z|^{2a} + |w|^{2b}\right)}\right) \\
&= \alpha_2 - \alpha_1.
\end{aligned}
\tag{24}
$$

Notice that the differential form α, defined by

$$
\begin{aligned}
\alpha = \alpha_1 &\quad \text{in} \quad U_1 \\
\alpha = \alpha_2 &\quad \text{in} \quad U_2
\end{aligned}
\tag{25}
$$

is well defined and smooth in all of $\mathbb{C}^2 - \{0\}$. It is also $\bar{\partial}$-closed. Computation shows that

$$
\alpha = \frac{\overline{a z^{a-1} w^b} d\bar{z} - \overline{b w^{b-1} z^a} d\bar{w}}{\left(|z|^{2a} + |w|^{2b}\right)^2}.
\tag{26}
$$

It is a standard exercise, using the extension theorem of Hartogs, to see that it is not $\bar{\partial}$-exact.

An alternative proof follows. If $\alpha = \bar{\partial}u$ for some smooth function u in the complement of the origin, then the $(2,1)$ form η defined by

$$
\eta = ab z^{a-1} w^{b-1} \alpha \wedge dz \wedge dw = d\left(u \, ab z^{a-1} w^{b-1} \, dz \wedge dw\right)
\tag{27}
$$

would be exact in $\mathbb{C}^2 - \{0\}$. Consider the domain

$$
\Omega = \left\{|z|^{2a} + |w|^{2b} < 1\right\}
\tag{28}
$$

and note that its boundary is a smooth compact manifold without boundary. Hence we would have $0 = \int_{b\Omega} \eta$. On the other hand, using the explicit form for α, the fact that the denominator is constant on the boundary, and Stokes' theorem, we obtain the following contradiction:

$$
\begin{aligned}
0 &= \int_{b\Omega} d\left(u\,abz^{a-1}w^{b-1}\,dz \wedge dw\right) = \int_{b\Omega} \alpha \wedge abz^{a-1}w^{b-1}\,dz \wedge dw \\
&= \int_{b\Omega} \left(a\overline{z^{a-1}w^b}\,d\bar{z} - b\overline{w^{b-1}z^a}\,d\bar{w}\right) abz^{a-1}w^{b-1}\,dz \wedge dw \\
&= \int_{\Omega} 2\left|abz^{a-1}w^{b-1}\right|^2 d\bar{w}\,d\bar{z}\,dz\,dw \neq 0.
\end{aligned}
\tag{29}
$$

Therefore α is not $\overline{\partial}$-closed, and the cohomology is therefore nontrivial. It is not difficult to verify that the differential forms for different choices of the integers (a, b) give independent members of the cohomology group, and hence that the cohomology is infinite dimensional. □

We use the term "residue integral" for any integral formula that computes the index of a finite analytic mapping about an isolated zero. Suppose that $(f) = (f_1, f_2, \ldots, f_n)$ is a finite analytic mapping. There are several integrals whose values equal the index of f. These include an integral of an $(n, 0)$ form, an integral of an $(n, n-1)$ form, and an integral of an (n, n) form. We focus on the first two for the moment. When $n = 1$, the first two integrals are the same. The denominator in the integral of the $(n, 0)$ form vanishes on the union of the zero sets of the components. Such residue integrals are consequently somewhat tricky. The Bochner–Martinelli kernel is the integral of an $(n, n-1)$ form over an object of dimension $2n - 1$. The denominator of the integrand equals a power of $\|f\|^2$ and hence vanishes at just one point. The idea behind the Bochner–Martinelli integral is essentially the same as the idea in Example 1. See [Kr, Ra] for more information on the Bochner–Martinelli integral, especially for its interpretation as the "fundamental solution operator" for the Cauchy–Riemann equations.

We now describe this integral. Suppose that $f : \Omega \subset \mathbb{C}^n \to \mathbb{C}^n$ is holomorphic, $f(0) = 0$, and that the origin is an isolated zero. Define a differential $(n, n-1)$ form on $\Omega - \{0\}$ as follows:

$$
K_f = c_n \frac{\sum_{j=1}^{n} (-1)^{j+1}\, \overline{f_j \partial f_1} \wedge \ldots \{\{\overline{\partial f_j}\}\} \wedge \ldots \overline{\partial f_n}}{\|f\|^{2n}} \partial f_1 \wedge \cdots \wedge \partial f_n.
\tag{30}
$$

The double braces around $\overline{\partial f_j}$ indicate that it is to be omitted in the exterior product. An elementary calculation shows that K_f is $\overline{\partial}$-closed. To see that it cannot be $\bar{\partial}$-exact, we mimic the method in Example 1, by computing the integral

$$\int_{\|f\|=\epsilon} K_f = c_n \epsilon^{-2n} \int_{\|f\|=\epsilon} \sum_{j=1}^{n} (-1)^{j+1} \overline{f_j \partial f_1} \wedge \ldots (\overline{\partial f_j}) \wedge \ldots \overline{\partial f_n} \, \partial f_1 \wedge \ldots \partial f_n$$

$$= c_n \epsilon^{-2n} \int_{\|f\|\leq\epsilon} n \left| \det(df) \right|^2 d\bar{z} \, dz. \tag{31}$$

This last expression does not vanish, because $\det(df)$ does not vanish identically. Corollary 2 makes a stronger assertion about the zero set of $\det(df)$.

To actually evaluate the integral, we choose the constant c_n so that (31) equals unity when f is the identity mapping. This requires

$$c_n = \frac{1}{n \int_B d\bar{z} \, dz} \tag{32}$$

where $B = B_n$ denotes the unit ball in $\mathbb{C}^n$. Note that the differential (n, n) form $d\bar{z} \, dz$ can be written

$$d\bar{z} \, dz = (-1)^{\frac{n(n+1)}{2}} (2i)^n \, dV \tag{33}$$

in terms of the usual volume form in real Euclidean space. Also note that $\text{Vol}(B_n) = \pi^n / n!$.

Let us record the result of this computation. The right side of (35) equals the index of the mapping. This follows from the change of variables formula (14) for a multiple integral. We also use Lemma 1.2 to know that $|\det(df)|^2 = \det(dF)$ if F is the underlying real mapping.

PROPOSITION 5
Suppose that

$$f : B_n \to \mathbb{C}^n \quad and \quad f^{-1}(0) = 0 \tag{34}$$

is a finite holomorphic mapping. Let K_f denote the Bochner–Martinelli $(n, n-1)$ form defined by (30). Then

$$\int_{\|f\|=\epsilon} K_f = \frac{n!}{\pi^n} \epsilon^{-2n} \int_{\|f\|\leq\epsilon} |\det(df)|^2 \, dV. \tag{35}$$

Either side of (35) equals the index of the mapping.

2.1.5 The Grothendieck residue integral

We have noted that the equidimensional finite analytic mappings behave like mappings in one variable. In particular they are open mappings and have finite index. In this section we give another possible definition of the multiplicity of such a mapping, using a residue integral. So, let $f : (\mathbb{C}^n, 0) \to (\mathbb{C}^n, 0)$ be (the germ of) a holomorphic mapping. Suppose f is defined on some open ball Ω about the origin, and it is a finite analytic mapping; thus $f^{-1}(0) = 0$. The mapping f does not vanish elsewhere on Ω, so there is a sphere about the origin within Ω on which

$$\| f(z) \| \geq c > 0. \tag{36}$$

Consider the real n-dimensional cycle $\Gamma = \Gamma_f$ that is defined by the equations $|f_j(z)| = \epsilon$ for sufficiently small positive ϵ, and is oriented so that

$$d\left(\arg(f_1)\right) \wedge \cdots \wedge d\left(\arg(f_n)\right) > 0. \tag{37}$$

A natural generalization of the one-variable residue integral is the integral

$$\mathbf{Res}(f) = \left(1/2\pi i\right)^n \int_{\Gamma} \frac{df_1}{f_1(z)} \wedge \frac{df_2}{f_2(z)} \wedge \cdots \wedge \frac{df_n}{f_n(z)} \tag{38}$$

which is independent of ϵ.

DEFINITION 3 *If $f : (\mathbb{C}^n, 0) \to (\mathbb{C}^n, 0)$ is the germ of a finite holomorphic mapping, then its (Grothendieck) residue about 0 is the integral $\mathbf{Res}(f)$.*

We will often abbreviate the differentials in the integrand. Recall the use of multi-index notation as in the Cauchy integral formula.

$$\frac{dz^1 \wedge \cdots \wedge dz^n}{f_1(z) \ldots f_n(z)} = \frac{dz}{f(z)}. \tag{39}$$

Introducing also the Jacobian determinant into the formula gives

$$\mathbf{Res}(f) = \left(1/2\pi i\right)^n \int_{\Gamma} \det\left(df(z)\right) \frac{dz}{f(z)}. \tag{40}$$

Note that the residue is an integral of an $(n, 0)$ form over an n-cycle. As mentioned in the previous section, it gives the same value as the Bochner–Martinelli integral, which is a surface integral of an $(n, n-1)$ form over a $(2n-1)$-dimensional surface. The value of the residue integral for a finite analytic mapping depends only on the ideal in the power series ring generated by the component functions. There is consequently no ambiguity in writing

$$\mathbf{Res}(f) = \mathbf{Res}(f_1, \ldots, f_n), \tag{41}$$

whether we conceive of the mapping f or the ideal (f) as the basic object of interest. See Theorem 1 for all these statements. See [GH] for a detailed discussion of this integral. In the next section we develop many of its properties and some efficient techniques for its computation. One can compute the integral

(38) in case $f(0)$ does not vanish; the result will be zero. If one computes it for a mapping that is not a finite analytic mapping, then the result will be infinity. Thus the number

$$\mathbf{Res}\,(f) = \mathbf{Res}\,(f_1, \ldots, f_n) \tag{42}$$

has these properties:

$$\mathbf{Res}\,(f) = 0 \Leftrightarrow \mathbf{V}\,(f) \text{ is empty}$$

$$0 < \mathbf{Res}\,(f) < \infty \Leftrightarrow \mathbf{V}\,(f) = \{0\}$$

$$\mathbf{Res}\,(f) = \infty \Leftrightarrow \dim \mathbf{V}\,(f) > 0. \tag{43}$$

Establishing these properties leads us toward our aim of measuring the singularity of a finite map germ. This number is not the only one with the properties (43), but it is the best one for many purposes. Roughly speaking, the reason is that it is stable, i.e., upper semicontinuous under perturbation. Other invariants will not be upper semicontinuous; see Example 6.

2.1.6 The codimension of the local algebra

In this section we discuss the codimension operation. We will follow the method in [AGV] in proving the equality of the index of a holomorphic map germ and its codimension. First we make the definition and some simple observations. Suppose $f : (\mathbb{C}^n, p) \to (\mathbb{C}^N, q)$ is the germ of a holomorphic mapping, and (f) denotes the ideal in $_nO_p$ generated by the component functions.

DEFINITION 4 *The local algebra of (f) at p is the quotient algebra $_nO_p/(f)$. The codimension of (f), written $\mathbf{D}(f)$, is the dimension as a complex vector space of the local algebra.*

$$\mathbf{D}\,(f) = \dim_{\mathbb{C}} \left(_nO_p/(f)\right). \tag{44}$$

Note that the codimension is defined for all ideals; let us generalize the trichotomy in the one-dimensional case we saw in Section 1.1. We work at the origin. Suppose that (f) is an ideal in O. We claim that the variety defined by (f) is positive dimensional if and only if the codimension is infinite. To see this, suppose that $\mathbf{D}(f)$ is finite, but not zero. Then every coordinate function lies in the radical, so $\mathrm{rad}\,(f) = \mathbf{M}$. Therefore $\mathbf{V}\,(f) = \mathbf{V}\,(\mathbf{M}) = \{0\}$. Conversely, if the variety consists of the origin alone, then the Nullstellensatz implies that $\mathrm{rad}\,(f) = \mathbf{M}$, so the codimension is finite. Thus $\mathbf{V}\,(f)$ is an isolated point if and only if $0 < \mathbf{D}(f) < \infty$. Also $\mathbf{V}\,(f)$ is positive dimensional if and only if $\mathbf{D}(f) = \infty$. Finally, the remaining trivial possibility is that $\mathbf{V}\,(f)$ is empty. This happens if and only if $(f) = O$, or equivalently, if and only if $\mathbf{D}(f) = 0$.

The preceding paragraph nicely generalizes to several variables the discussion in one dimension. The main point is that f is a finite analytic mapping if and only if the codimension is finite and nonzero. Theorem 1 shows that, in the equidimensional case, the codimension equals precisely the index. The codimension is defined more generally and has additional interpretations. Suppose f is a finite analytic mapping, and h vanishes at the origin. The Nullstellensatz guarantees that some power of h lies in (f). One can always take this power to be $\mathbf{D}(f)$. This property is statement 8 of Theorem 1 in the next section, but we derive it now.

By the Nullstellensatz, we know that if $\mathbf{V}(f)$ consists of the origin alone, and if h vanishes at the origin, then h lies in rad(f). Let K be the smallest integer for which $h^K \equiv 0 \bmod (f)$. It must be shown that $K \leq \mathbf{D}(f)$. Suppose this were false. If $K > \mathbf{D}(f)$, then the powers of h, namely $1, h, h^2, \ldots, h^D$ would not lie in the ideal. The desired contradiction follows from the linear independence of these powers.

To verify this, suppose that there is a nontrivial relationship

$$0 \equiv \sum_{j=0}^{D} c_j h^j \bmod (f). \tag{45}$$

Let k denote the smallest integer for which $c_k \neq 0$. Then we can write

$$0 \equiv \sum_{j=k}^{D} c_j h^j \equiv h^k (c_k + \cdots) \equiv u h^k \bmod (f) \tag{46}$$

for the unit $u = c_k + \ldots$. This then implies that

$$0 \equiv h^k \bmod (f). \tag{47}$$

Since $k \leq D$, we are contradicting the definition of K.

The next observation is that the equality of the codimension and the index is obvious for the ideals

$$(f) = (z^m) = \left(z_1^{m_1}, \ldots, z_n^{m_n} \right) \tag{48}$$

as both numbers equal $\prod_{j=1}^{n} m_j$. On the other hand, suppose that (f) is any ideal for which $\mathbf{D}(f)$ is finite. Then, the ideal $(z^m + \epsilon f)$ defines the same ideal as (f) if the exponents are chosen large enough. This follows, since we have just proved that $z_j^{m_j}$ lies in the ideal (f), if we take $m_j = \mathbf{D}(f)$. Fix now a small neighborhood of the origin, and let $\{p_i(\epsilon)\}$ denote the roots of the mapping

$$f_\epsilon(z) = \epsilon f(z) + z^m. \tag{49}$$

We will need to consider the codimension defined at each of these points. Therefore write g_p for the germ of g at p and $\mathbf{D}_p(g) = \dim(O_p/g_p)$ for the codimension there. Let us assume for the moment the inequalities

$$\mathbf{D}(f) \geq \sum_i \mathbf{D}_{p_i}(f_\epsilon)$$

$$\mathbf{D}_{p_i}(g) \geq \text{index}_{p_i}(g). \tag{50}$$

These are Lemmas 1 and 2 below. We now proceed to the proof. Combining (50) with the coalescing property (15) of the index gives

$$\mathbf{D}(z^m) \geq \sum_i \mathbf{D}_{p_i}(z^m + \epsilon f)$$

$$\geq \sum_i \text{index}_{p_i}(z^m + \epsilon f)$$

$$= \text{index}_o(z^m). \tag{51}$$

Since equality of the codimension and the index hold for the mapping (z^m), we conclude that all the inequalities used in (51) are actually equalities. In particular, at each p_i, we must have

$$\mathbf{D}_{p_i}(z^m + \epsilon f) = \text{index}_{p_i}(z^m + \epsilon f). \tag{52}$$

The origin is one of these points, so equality holds there also. Therefore

$$\mathbf{D}(f) = \mathbf{D}_o(f) = \mathbf{D}_o(z^m + \epsilon f)$$

$$= \text{index}_o(z^m + \epsilon f) = \text{index}_o(f), \tag{53}$$

which is what we are trying to prove. Thus the equality of the index and the codimension now follows from the following two lemmas.

LEMMA 1
If (f_ϵ) is a continuous deformation of (f), and $\{p_i(\epsilon)\}$ denotes the roots of the perturbation, then

$$\mathbf{D}(f) \geq \sum_i \mathbf{D}_{p_i}(f_\epsilon). \tag{54}$$

Suppose that the perturbed mapping still has the origin as a root. Then, in particular, taking codimensions at the origin yields $\mathbf{D}(f) \geq \mathbf{D}(f_\epsilon)$. This is so because, if a function does not lie in an ideal, then disturbing its coefficients or the coefficients of the generators of the ideal by a sufficiently small amount will not make it be in the ideal. Therefore the dimension of the quotient space cannot decrease upon taking limits of parameter values. Since it is integer-valued, the codimension must be upper semicontinuous. The most trivial example is $f_\epsilon(z) = \epsilon z + z^2$ in one variable. Working at the origin, we obtain $\mathbf{D}(f) = 2$ while $\mathbf{D}(f_\epsilon) = 1$.

LEMMA 2

The codimension is not smaller than the index, that is,

$$\mathbf{D}_o(f) \geq \text{index}_0(f). \tag{55}$$

The proofs of these lemmas are important steps in the proof of Theorem 1. Detailed proofs of them appear in [AGV]. ∎

REMARK 1 According to Theorem 1, the codimension is but one of several equivalent definitions of the multiplicity of the singularity. It is particularly important because it does not rely on the equidimensionality of the mapping. Consider an ideal that defines a zero-dimensional variety, but is not a complete intersection. A complete intersection is a variety (together with its defining equations) whose dimension equals $n - N$, the number of variables minus the number of defining equations. In order to use the methods of residue integrals or degrees of mappings, one must introduce extra variables as parameters. One can compute the codimension directly. ∎

REMARK 2 Consider the mapping defined by $f(z) = (z^2, z^3)$ where the superscripts are exponents, and $z \in \mathbb{C}$. If we consider this mapping as taking values in the complex analytic subvariety $\{z_1^3 = z_2^2\}$ of $\mathbb{C}^2$, then it is a homeomorphism. Therefore its "index" should be unity. On the other hand, if we consider it as taking values in $\mathbb{C}^2$, then it has a singularity at the origin. The codimension operator assigns this mapping the (more reasonable) number two. See Definition 7 where we define the "multiplicity" or "order" of a parameterized holomorphic curve. This remark indicates why Theorem 1 applies only for equidimensional mappings. ∎

Example 2

We suppose $n = 2$ and write the variables as (z, w). Consider the ideal

$$(f) = (z^3, zw, w^3). \tag{56}$$

Computation shows that a basis for $O/(f)$ is given by

$$\{1, z, z^2, w, w^2\}. \tag{57}$$

Consider a new variable, denoted ζ, and the ideal in three dimensions given by

$$(f_\zeta) = (z^3, w^3, zw - \zeta). \tag{58}$$

Computation shows that a basis for $_3O/(f_\zeta)$ is given by

$$\{1, z, z^2, w, wz, wz^2, w^2, w^2z, w^2z^2\} = \{1, z, z^2, w, \zeta, \zeta z, w^2, w\zeta, \zeta^2\}. \tag{59}$$

Thus the quotient space is nine-dimensional. Taking the limit as ζ tends to zero, we obtain the five nonzero elements in $O/(f)$. At first glance this seems to contradict the upper semicontinuity of the codimension operator. It does not, because $\dim\left({}_3O/(f)\right) = \infty$ while

$$\mathbf{D}(f) = \dim\left({}_2O/(f)\right) = 5. \tag{60}$$

See Section 3.5 for more computations of the codimension.   ▯

2.2   Intersection numbers

2.2.1   Equivalence of various definitions

In this section we discuss in detail the various definitions of the winding number of an equidimensional mapping. Recall that one definition is given in terms of a residue integral:

$$\mathbf{Res}(f) = \left(1/2\pi i\right)^n \int_\Gamma \det\left(df(z)\right) \frac{dz}{f(z)} \,. \tag{61}$$

It is clear from Theorem 1 below that there are many possible equivalent definitions, each with its own point of view. The history of algebraic geometry reveals two predominant points of view on these matters, the static and the dynamic. In the "static" method, one computes intersection numbers by staying at the point in question and doing algebra. In the "dynamic" method, one passes to nearby points. See [Fu] for references. It is difficult to prove even the "static" statements of Theorem 1 without "dynamic" ideas.

THEOREM 1
Suppose that $f : (\mathbb{C}^n, 0) \to (\mathbb{C}^n, 0)$ is a finite analytic mapping. The intersection number $\mathbf{Res}(f)$ has the following properties:

1. $\mathbf{Res}(f)$ *is a positive integer.*
2. $\mathbf{Res}(f)$ *depends only on the ideal (f) in ${}_nO$ and not on the choice of generators.*
3. $\mathbf{Res}(f)$ *is the generic number of solutions to the equation $f(z) = w$, for w sufficiently close to the origin.*
4. $\mathbf{Res}(f)$ *is the topological degree of f, that is,*

$$\mathbf{Res}(f) = \mathrm{index}_0(f). \tag{62}$$

5. *The residue integral equals the integral of the Bochner–Martinelli kernel:*

$$\mathbf{Res}\,(f) = \int\limits_{\|f\|=\epsilon} K_f. \tag{63}$$

6. *The residue integral equals the codimension of the ideal:*

$$\mathbf{Res}\,(f) = \mathbf{D}\,(f) = \dim_{\mathbb{C}} \left(O/\left(f \right) \right). \tag{64}$$

7. *Put* $f' = \left(f_1, \ldots, f_{n-1} \right)$. *Suppose that* $\left(f', gh \right)$ *is a finite analytic mapping. Then*

$$\mathbf{D}\left(f', gh \right) = \mathbf{D}\left(f', g \right) + \mathbf{D}\left(f', h \right). \tag{65}$$

8. *If* h *is an arbitrary element of the maximal ideal* **M**, *then* $h \in \mathrm{rad}\,(f)$, *and*

$$h^{\mathbf{D}(f)} \in (f). \tag{66}$$

9. *If the Taylor coefficients of* f *depend continuously on some parameters, then* $\mathbf{Res}\,(f)$ *depends upper semicontinuously on them. In other words,*

$$\mathbf{Res}\,(f_\epsilon) \leq \mathbf{Res}\,(f) \tag{67}$$

if f_ϵ *is a sufficiently small perturbation of* f.

All of the statements in Theorem 1 are important. We will use each property and we will require an additional formula (Theorem 3) that relates the codimension to the order of vanishing of certain pullbacks to parameterized holomorphic curves. This formula is a special case of a deep theory. Intersection theory is highly developed; methods of homological algebra that are beyond the scope of this book play a leading role (see [Fu]).

We have proved already many of the statements in Theorem 1, so for clarity we summarize our work thus far. We have discussed the equality of the codimension and the index for finite analytic mappings. Either of these numbers is obviously a positive integer, and the codimension by definition depends only on the ideal. Proposition 5 shows that the index equals the Bochner–Martinelli integral. Thus statements 1 through 4 and statement 6 all follow from statement 5. So does statement 9, since we have verified the upper semicontinuity property of the codimension operator. Statements 7 and 8 also involve properties of the codimension; statement 8 appeared in Section 2.1.6, while statement 7 will be proved in Section 2.2.3 on regular sequences. It remains to establish statement 5, namely, the equality of the residue integral and the Bochner–Martinelli integral.

PROOF OF A SPECIAL CASE OF STATEMENT 5   In order to prove statement 5, we could appeal to cohomology theory. The equality of the two integrals is a consequence of simple facts from Dolbeault cohomology. We prove it only in the two-dimensional case, using elementary ideas. The technique amounts to iterating the proof of Proposition 5. Recall that the Bochner–Martinelli kernel is given by

$$K_f = c_n \sum_{j=1}^{n} (-1)^{j+1} \frac{\overline{f_j}}{||f||^{2n}} \overline{\partial f_1} \wedge \ldots \{\{\overline{\partial f_j}\}\} \ldots \wedge \overline{\partial f_n} \wedge \partial f_1 \wedge \ldots \partial f_n. \quad (68)$$

In the formula for $K(f)$, recall that the differential $\overline{\partial f_j}$ is missing from the wedge product, and the constant is chosen so that

$$1 = nc_n \int_{B_n} d\bar{z}\, dz. \quad (69)$$

We need the equality that

$$\mathbf{Res}\,(f) = \left(\frac{1}{2\pi i}\right)^n \int_{\Gamma} \frac{\det(df)}{f}\, dz = \int_{||f||=\epsilon} K_f. \quad (70)$$

Here is the proof in the two-dimensional case:

The idea is the same as in the proof that there are $\overline{\partial}$-closed forms that are not $\overline{\partial}$-exact in the complement of the origin. Multiplying and dividing by $||f||^2 = |f_1|^2 + |f_2|^2$ yields the following:

$$\int_{|f_1|=\epsilon} \int_{|f_2|=\epsilon} \frac{(\det(df))}{f_1 f_2}\, dz_1\, dz_2 = \int_{|f_1|=\epsilon} \int_{|f_2|=\epsilon} \frac{\left(|f_1|^2 + |f_2|^2\right)(\det(df))}{||f||^2 f_1 f_2}\, dz_1\, dz_2$$

$$= \int_{|f_1|=\epsilon} \int_{|f_2|=\epsilon} \frac{\overline{f_1}\,(\det(df))}{||f||^2 f_2} + \frac{\overline{f_2}\,(\det(df))}{||f||^2 f_1}\, dz_1\, dz_2$$

$$= \int_{|f_1|=\epsilon} \int_{|f_2|=\epsilon} \frac{\overline{f_1}\,(\det(df))}{||f||^2 f_2}\, dz_1\, dz_2$$

$$+ \int_{|f_1|=\epsilon} \int_{|f_2|=\epsilon} \frac{\overline{f_2}\,(\det(df))}{||f||^2 f_1}\, dz_1\, dz_2. \quad (71)$$

Now notice that each integrand is singular on a smaller set than was the original integrand. In each case we can apply Stokes' theorem as above. Keeping in mind the orientation, we obtain for (71)

$$- \int_{|f_2|=\epsilon} \int_{|f_1|\leq\epsilon} \overline{\partial}\left(\frac{\overline{f_1}\,(\det(df))}{||f||^2 f_2}\, dz_1\, dz_2\right)$$

$$+ \int_{|f_1|=\epsilon} \int_{|f_2|\leq\epsilon} \overline{\partial}\left(\frac{\overline{f_2}\,(\det(df))}{||f||^2 f_1}\, dz_1\, dz_2\right). \quad (72)$$

Computing the indicated $\overline{\partial}$ derivatives by calculus, we obtain

$$- \int\limits_{|f_2|=\epsilon} \int\limits_{|f_1|\le\epsilon} \frac{\overline{f_2\partial f_1} - \overline{f_1\partial f_2}}{\|f\|^4} \det(df)\, dz_1\, dz_2$$

$$+ \int\limits_{|f_1|=\epsilon} \int\limits_{|f_2|\le\epsilon} \frac{\overline{f_1\partial f_2} - \overline{f_2\partial f_1}}{\|f\|^4} \det(df)\, dz_1\, dz_2. \qquad (73)$$

Noting that each integrand is essentially the Bochner–Martinelli kernel, and putting in the factor $(1/2\pi i)^2$, we see that

$$\mathbf{Res}\,(f) = \frac{1}{2}\int\limits_A K_f + \frac{1}{2}\int\limits_{A'} K_f \qquad (74)$$

where the sets of integration are given by

$$\{|f_1| \le \epsilon, |f_2| = \epsilon\} \quad \text{and} \quad \{|f_1| \le \epsilon, |f_2| = \epsilon\} \qquad (75)$$

respectively. On the other hand, by another application of Stokes' theorem, we can replace each set of integration by the set $\|f\|^2 \le \epsilon^2$ because the Bochner–Martinelli form is closed in any region not containing a root. This finishes the proof in two dimensions. The general case requires an iteration of this procedure and appears in [GH]. ∎

Combining statement 5 with Proposition 5 gives us three equal integrals

$$\left(\frac{1}{2\pi i}\right)^n \int\limits_\Gamma \frac{\det(df)}{f}\, dz = \int\limits_{\|f\|=\epsilon} K_f = \frac{\epsilon^{-2n}}{\mathrm{Vol}\,(B_n)} \int\limits_{\|f\|\le\epsilon} |\det(df)|^2\, dV. \qquad (76)$$

The third integral gives a formula for the residue integral that involves no singularities. This integral equals the topological degree of the mapping, because the integrand is the Jacobian determinant of the underlying real mapping.

According to (76) the residue equals the topological degree of the mapping. As we have noted, a holomorphic mapping is orientation preserving, so the topological degree equals the generic number of roots to the equation $f(z) = w$ for w sufficiently close to the origin.

It is useful to remark that the Bochner–Martinelli kernel gives an explicit way of seeing the method of perturbation. Consider the case when f has finitely many zeroes inside Γ. Let S_ϵ be a sufficiently small sphere that encloses all these points but no other zeroes of f. Consider residue integrals about each of them and find their sum. The result equals the Bochner–Martinelli integral around S_ϵ. This follows immediately from Stokes' theorem because K_f is closed. The equality of these integrals enables us to perturb f so as to let the multiple zeroes come apart.

2.2.2   Consequences of Theorem 1

Theorem 1 has numerous corollaries. These results help develop our point of view. One key idea is that equidimensional finite analytic mappings behave essentially as nonconstant mappings in one variable do. This holds in the algebraic, geometric, and analytic senses and affords us thereby a valuable intuition about such mappings. In this section we offer two other corollaries, each of which arises in the study of germs of holomorphic (or even CR) mappings in several variables.

First we derive a nice fact about the Jacobian determinant. To begin the discussion, consider the (more general) residue integral

$$\left(\frac{1}{2\pi i}\right)^n \int_\Gamma \frac{h(z)}{f(z)}\, dz \tag{77}$$

where h is an arbitrary element of O. By our conventions on multi-indices, the denominator represents the product $\prod_{j=1}^n f_j$ of the component functions. The number $\mathbf{Res}(f)$ results from choosing $h = \det(df)$. The following lemma is the easy half of an important theorem of Grothendieck.

LEMMA 3
Suppose that h lies in the ideal (f). Then the residue integral vanishes:

$$\left(\frac{1}{2\pi i}\right)^n \int_\Gamma \frac{h(z)}{f(z)}\, dz = 0. \tag{78}$$

PROOF   Any element of the ideal can be written $h = \sum_{j=1}^n a_j f_j$. By the linearity of the integral it is sufficient to prove the assertion in case there is just one term, say $h = a_n f_n$. We must then verify that

$$\left(\frac{1}{2\pi i}\right)^n \int_\Gamma \frac{a_n(z) f_n(z)}{f_1(z) \dots f_n(z)}\, dz = 0. \tag{79}$$

After cancellation, we no longer divide by f_n. Thus the integrand is now holomorphic in the complement of $\cup_{j=1}^{n-1} \mathbf{V}(f_j)$. Thus we can apply Stokes' theorem to this integral of an $(n, 0)$ form over an n-boundary to obtain an integral of an $(n, 1)$ form over an $n + 1$ cycle. Since the form is holomorphic, it is closed, and we obtain the desired result:

$$\int_{|f_1|=\epsilon} \cdots \int_{|f_{n-1}|=\epsilon} \int_{|f_n|\leq\epsilon} d\left(\frac{a_n(z)}{f_1(z) \dots f_{n-1}(z)}\, dz\right) = 0. \quad \blacksquare \tag{80}$$

The other half of Grothendieck's result is much deeper. It states that h lies in the ideal (f) only if, for every g, the residue integral vanishes:

$$\int_\Gamma \frac{hg}{f} \, dz = 0. \tag{81}$$

See [AGV,GH] for proofs. It is worth observing that, even in one variable, the knowledge that (81) holds for g equal to unity is not sufficient for the conclusion.

COROLLARY 2
If f is a finite holomorphic map germ, then

$$\det(df) \not\equiv 0 \bmod (f). \tag{82}$$

PROOF According to the theorem,

$$\left(\frac{1}{2\pi i}\right)^n \int_\Gamma \frac{\det(df)}{f} \, dz = \dim\left(O/(f)\right) \neq 0 \tag{83}$$

so Lemma 3 implies that $\det(df) \not\equiv 0 \bmod (f)$. ∎

Algebraists know considerably more about these ideas. One piece of information is that, although the Jacobian $\det(df)$ is not zero modulo (f), $h\det(df) \equiv 0$ for every element h vanishing at the origin. Thus $\det(df)$ is, in some sense, as close to zero as possible. There is also a formula relating the Jacobian to the codimension. Since $f(0) = 0$, we can write $f(z) = A(z)\,z$ for some holomorphic matrix $A(z)$. Then we have the formula that

$$\mathbf{D}(f)\det(A) \equiv \det(df) \bmod (f). \tag{84}$$

The reader can derive this formula from the Grothendieck result and the change of variables formula in an integral. The author believes that this formula should have an analytic application in the study (see Chapter 6) of subelliptic estimates, but he has not yet been able to find one.

The second corollary concerns the behavior of the codimension operator under finite mappings. Suppose that $f : (\mathbb{C}^n, 0) \to (\mathbb{C}^n, 0)$ is the germ of a finite holomorphic mapping with multiplicity $\mathbf{D}(f)$. If $I \subset O_n$ is a proper ideal, then we define $f^*(I)$ to be the ideal generated by all germs of the form $f^*g = g(f)$ for $g \in I$. The proof of the following result then follows from the interpretation of the codimension as the index of the mapping.

COROLLARY 3
Under the hypotheses of the preceding paragraph, and supposing also that $\mathbf{D}(I)$ is finite, it follows that $\mathbf{D}(f^*I)$ is finite and satisfies

$$\mathbf{D}\left(f^*I\right) = \mathbf{D}(f)\mathbf{D}(I). \tag{85}$$

Let us discuss here an alternative approach to statement 8 of Theorem 1 that does not use the Nullstellensatz. We have shown already that, for w close enough to 0, the system of equations $f(z) = w$ has $m = \mathbf{D}(f)$ solutions, counting multiplicities. Call them $z_j(w)$. Consider, for any holomorphic h with $h(0) = 0$, the polynomial in h defined by

$$H(z, w) = \prod_{j=1}^{m} \left(h(z) - h\left(z_j(w) \right) \right). \tag{86}$$

It is obvious that $H(z, f(z)) = 0$ because the $z_j(w)$ are solutions. On the other hand, even though these roots are not holomorphic functions of z, the manners in which they occur in the expanded form of $H(z, f(z))$ are holomorphic. This is similar to the phenomenon that occurs in the proof of the preparation theorem. See also the proof of the Nullstellensatz. Solving $H(z, f(z)) = 0$ for its highest order term yields $h(z)^m = \sum_{j=1}^{n} g_j(z) f_j(z)$ for appropriate g_j, and hence h^m lies in the ideal.

We illustrate this for an ideal generated by quadratic polynomials. For ease on the eyes, write $z = (a, b)$, $w = (c, d)$. Suppose that

$$f(a, b) = \left(a^2 - b^2, ab \right). \tag{87}$$

We claim that $a^4 \in (f)$. This is trivial to check by writing down that

$$a^4 = a^2 \left(a^2 - b^2 \right) + (ab)(ab). \tag{88}$$

The interpretation using (86) yields (88) directly. Solve the equations (87) for (a, b) in terms of (c, d) to obtain that a satisfies

$$a^2 - \left(\frac{d}{a} \right)^2 = c. \tag{89}$$

Multiply through by a^2 and subtract. This yields the polynomial $H(z, w) \simeq H(a, b, c, d)$ that corresponds to the polynomial $h(z) \simeq h(a, b) = a$. We obtain (88) and thereby verify that a^4 is an element of (f).

REMARK 3   The special case of the Nullstellensatz for zero-dimensional varieties is somewhat more general; here we are assuming that the ideal is generated by the same number of functions as there are variables. This situation, called a complete intersection of dimension zero, or a regular sequence of order n, is somewhat simpler than the general case. We discuss regular sequences in Section 2.4. It is possible to reduce some problems to the complete intersection case by introducing additional parameters.

Given an ideal that defines a zero-dimensional variety, a generic choice of n linear combinations among the generators also defines a zero-dimensional variety, and hence is a complete intersection. Writing $\mathbf{D}(f)$ now for the codimension of the ideal, i.e.,

$$\mathbf{D}(f) = \dim_{\mathbb{C}} O/(f), \tag{90}$$

we see that

$$\mathbf{D}(f_1, \ldots, f_N) \leq \mathbf{D}(g_1, \ldots, g_n) \tag{91}$$

if the g_i are generic. ∎

It was noted in Theorem 1 that every monomial of sufficiently high degree also lies in the ideal. Thus it makes sense to measure the singularity of the mapping, or for any ideal, by the following number.

DEFINITION 5 *Let $f : (\mathbb{C}^n, 0) \to (\mathbb{C}^N, 0)$ be the germ of a finite holomorphic mapping. We define*

$$\mathbf{K}(f) = \inf \left\{ k : \mathbf{M}^k \subset (f) \right\}. \tag{92}$$

This measurement differs from the intersection number, but the following inequalities indicate the sense in which they are essentially equivalent.

PROPOSITION 6
For the germ of a finite holomorphic mapping $f : (\mathbb{C}^n, 0) \to (\mathbb{C}^N, 0)$, we have

$$\mathbf{K}(f) \leq \mathbf{D}(f)$$
$$\mathbf{D}(f) \leq \begin{pmatrix} \mathbf{K}(f) + n - 1 \\ \mathbf{K}(f) - 1 \end{pmatrix}. \tag{93}$$

The proof of this proposition will appear as part of a more general set of inequalities in Theorem 4.

2.2.3 Regular sequences

There is an algebraic sense in which k independent equations in n space define a variety of dimension $n - k$. Basic to this idea is the notion of a regular sequence. In this section we discuss enough of these ideas so that we can prove (essentially following [Sfv]) the remaining part of Theorem 1, namely statement 7.

Suppose that Λ is commutative local ring with identity and maximal ideal M. Examples to be kept in mind are the power series ring $_nO$ and certain quotients. Suppose that

$$f = (f_1, f_2, \ldots, f_k) \tag{94}$$

is an ordered collection of elements of the maximal ideal M. In the special case where these are holomorphic functions, defined near 0 in $\mathbb{C}^n$, this means that $f_j(0) = 0 \quad \forall j$.

DEFINITION 6 *Let Λ be a commutative local ring with identity and maximal ideal M. The ordered collection*

$$f = (f_1, f_2, \ldots, f_k) \tag{95}$$

of elements of M is called a regular sequence of order k, if

$$f_{j+1} \qquad j = 0, 1, 2, \ldots, k - 1 \tag{96}$$

is not a zero divisor in the ring $\Lambda / (f_1, \ldots, f_j)$.

Note that the definition assumes that the elements are in order. It is not obvious that the property of being a regular sequence is independent of the ordering. This statement does hold in a local ring. One can make the definition in other rings, such as the polynomial ring, but independence of the ordering may fail.

Example 3

Let $\Lambda = \mathbb{C}[a, b, c]$ be the polynomial ring in three variables, and let

$$f = (a, b(1 - a), c(1 - a)). \tag{97}$$

Then f satisfies (96) for k equal to three, but does not upon a reordering of the components; for, if

$$g = (b(1 - a), c(1 - a)), \tag{98}$$

then the second term is a divisor of zero in the quotient by the first.   []

LEMMA 4

The property that $f = (f_1, \ldots, f_n)$ is a regular sequence in a local ring is unchanged upon a reordering of the indices.

PROOF It is enough to show that the statement is valid if we switch two adjacent indices, because these switches generate all permutations. Suppose that these indices are $k + 1$ and $k + 2$. By working in the quotient ring,

$$\Lambda / (f_1, \ldots, f_k), \tag{99}$$

we see that it is enough to prove the lemma for pairs of elements (f, g). Thus we suppose that f is not a zero divisor in a local ring Λ, and that g is not a zero divisor in $\Lambda / (f)$. We must now prove that the same statements hold with f and g interchanged. Suppose there exists an element h such that $gh = 0$. We claim that h must vanish. This claim implies the first statement. To verify the claim, we show that h "vanishes to infinite order"; more precisely, we show that

$$h \epsilon (f)^m \quad \forall m. \tag{100}$$

It then follows from standard commutative algebra [Sfv] that h must vanish. Thus statement (100) is true for $m = 0$ and follows for all m by induction. Namely, supposing that

$$h = wf^m \quad \text{and} \quad gh = 0 \tag{101}$$

we see that $gwf^m = 0$. Since f is not a zero divisor, $gw = 0$. Since g is not a zero divisor mod (f), w must vanish mod (f), i.e., $w = kf$. This implies that $h \in (f^{m+1})$, completing the induction. Once we know that g is not a zero divisor, we must show that it cannot happen that fp is divisible by g for any p. This condition means that $fp = gq$, so the result follows because g is not a zero divisor mod (f).   ∎

PROPOSITION 7
Suppose $f = (f_1, \ldots, f_n)$ is a collection of germs of holomorphic functions vanishing at 0, and $f : (\mathbb{C}^n, 0) \to (\mathbb{C}^n, 0)$ is a finite analytic mapping. Then $f = (f_1, \ldots, f_n)$ is a regular sequence of order n in $_n O$. It follows that the dimension of the variety

$$\mathbf{V}\left(f_1, \ldots, f_j\right) \tag{102}$$

is exactly $n - j$.

PROOF We proceed by induction on the dimension. The one-dimensional case is easy. Since f_1 is a finite mapping, there is an integer m and a unit u so that $f_1(z) = z^m u(z)$. Thus f_1 is not a divisor of zero, and the result holds.

Suppose now that we have proved in dimension $n - 1$ that the components of any equidimensional finite mapping define a regular sequence. Let $f = (f_1, \ldots, f_n)$ denote our given mapping. We may assume that $\dim \mathbf{V}\left(f_1, \ldots, f_{n-1}\right) = 1$. Therefore, for a generic linear function h, we have

$$\mathbf{V}\left(f_1, \ldots, f_{n-1}, h\right) = \{0\}, \tag{103}$$

so this new mapping is a finite analytic mapping. Working in the quotient $_n O / (h)$, which we may identify with $_{n-1} O$, we see that the residue classes of the components of f form a regular sequence by the inductive hypothesis. This means that

$$\left(h, f_1, f_2, \ldots, f_{n-1}\right) \tag{104}$$

is a regular sequence in $_n O$. According to Lemma 4, rearranging the generators preserves this property, so we discover that

$$\left(f_1, \ldots, f_{n-1}, h\right) \tag{105}$$

is also a regular sequence. Now we must show that f_n is not a zero divisor in

$$_n O / \left(f_1, \ldots, f_{n-1}\right). \tag{106}$$

By the Nullstellensatz, we can write for some exponent

$$h^m = \sum_{j=1}^{n} g_j f_j \equiv g_n f_n \quad \text{in} \quad {}_nO/\left(f_1, \ldots, f_{n-1}\right). \tag{107}$$

This shows that if f_n were a zero divisor, then h^m and hence h would be also. We have already proved, however, that

$$\left(f_1, \ldots, f_{n-1}, h\right) \tag{108}$$

is a regular sequence. Thus f_n is not a zero divisor and the proof is finished. ∎

COMPLETION OF PROOF OF THEOREM 1   Let us now derive the additivity property that was statement 7 of Theorem 1. Suppose that $\left(f_1, \ldots, f_{n-1}, gh\right)$ is a regular sequence of order n. Then we wish to show that

$$\mathbf{D}\left(f_1, \ldots, f_{n-1}, gh\right) = \mathbf{D}\left(f_1, \ldots, f_{n-1}, g\right) + \mathbf{D}\left(f_1, \ldots, f_{n-1}, h\right). \tag{109}$$

To prove this we let A denote the ring $O/\left(f_1, \ldots, f_{n-1}\right)$. Denote also by g, h their residue classes in A. In this notation (109) becomes

$$\dim\left(A/(gh)\right) = \dim\left(A/(g)\right) + \dim\left(A/(h)\right). \tag{110}$$

Because $\left(f_1, \ldots, f_{n-1}, gh\right)$ is assumed to be a regular sequence, gh is not a zero divisor in A. Therefore neither is g. This implies that $A/(h)$ is isomorphic to $(g)/(gh)$ as a vector space and hence

$$\dim\left(A/(h)\right) = \dim\left((g)/(gh)\right). \tag{111}$$

Write

$$A/(gh) = A/(g) + (g)/(gh) \tag{112}$$

as a direct sum, and take dimensions. Since the dimension of a vector space is the sum of the dimensions of its direct summands, (112) and (111) imply (110). This completes the proof of statement 7 of Theorem 1; completing this proof finishes the proof of that theorem. ∎

Formula (109) has an important application that we can only mention. The notion of intersection number extends to meromorphic functions. Assuming

(109) remains valid in that case, we are led to the definition that

$$\mathbf{D}\left(f_1, \ldots, f_{n-1}, \frac{g}{h}\right) = \mathbf{D}\left(f_1, \ldots, f_{n-1}, g\right) - \mathbf{D}\left(f_1, \ldots, f_{n-1}, h\right). \quad (113)$$

The additivity extends similarly to each slot. Notice that (113) agrees with the holomorphic case. If $h = 1$, then $D\left(f_1, \ldots, f_{n-1}, 1\right) = 0$, because the ideal is then the whole ring. Once one has the formula (113), one can apply it to projective varieties and obtain such consequences as Bezout's theorem. See [Fu,Sfv].

2.3 The order of contact of an ideal

2.3.1 Definition and basic properties

The focus of this book is the interplay between real and complex objects. Of particular interest is whether a given real subvariety contains nontrivial complex subvarieties. The importance of this question is apparent already in Theorem 1.11, and it will become more apparent as we proceed.

In elementary geometry one decides whether a particular curve lies in a given surface by substituting parametric equations for the curve into the defining equation of the surface. This simple idea is the basis for the present section. Given an ideal $I \subset O$ we define an invariant, called the order of contact of I, by pulling back its elements to all possible nonconstant parameterized holomorphic curves. In Chapter 3 we discuss the analogous invariant for ideals of real-valued functions. The author suspects that many results in analysis on pseudoconvex domains will reach eventually a simple exposition by use of these ideas.

In order to give the definition of the order of contact of an ideal in O, we recall the notion of parameterized holomorphic curve and its multiplicity. In the proof of Corollary 3 of Chapter 1 we proved and used the fact that the sphere contains no parameterized holomorphic curves.

DEFINITION 7 *A parameterized holomorphic curve is a nonconstant holomorphic mapping*

$$z : U \to \mathbb{C}^n \quad (114)$$

where U is an open subset of $\mathbb{C}$. Let

$$z : (\mathbb{C}, 0) \to (\mathbb{C}^n, 0) \quad (115)$$

be the germ of a parameterized holomorphic curve. Since z is not a constant, there is a least integer $v = v(z) = v_0(z)$ for which some derivative of z of order v does not vanish at 0. This integer is called the multiplicity, or order, of z at 0.

We use the term "parameterized holomorphic curve" for both the mapping and its image in $\mathbb{C}^n$. If we write a parameterized holomorphic curve z in terms of components, so

$$z = \left(z^1, \ldots, z^n \right), \tag{116}$$

then it is clear that $v(z) = \min \left\{ v\left(z^j\right) \right\}$. We call the curve smooth, or regular, at 0, if it has multiplicity equal to one there. Note also that the multiplicity of the curve z is unchanged if we make biholomorphic changes of coordinates in the domain or range that preserve the respective origins.

Suppose now that I is a proper ideal in the ring O of germs of holomorphic functions at 0 in $\mathbb{C}^n$, and z is the germ of a holomorphic curve. We consider the rational number

$$\mathbf{T}(I, z) = \inf_{g \in I} \frac{v(z^* g)}{v(z)} \tag{117}$$

for each nonconstant germ of a holomorphic curve.

DEFINITION 8   *The order of contact of a proper ideal I in O is defined to be*

$$\mathbf{T}(I) = \sup_z \left(\mathbf{T}(I, z) \right) = \sup_z \inf_{g \in I} \frac{v(z^* g)}{v(z)} . \tag{118}$$

Some remarks amplify the definition. The supremum is taken over all nonconstant parameterized holomorphic curves z with $z(0) = 0$, and the infimum is taken over all elements of the ideal. The number $\mathbf{T}(I)$ is unchanged if we make a local biholomorphic change of coordinates in either $\mathbb{C}$ or $\mathbb{C}^n$. If we replace the variable t in $\mathbb{C}$ by some nonconstant holomorphic function $h(t)$ preserving the origin, then both the numerator and denominator of

$$\frac{v(z^* g)}{v(z)} \tag{119}$$

get multiplied by the same factor, namely $v(h)$. As a consequence, $\mathbf{T}(I, z)$ depends only on the image of the curve and not on its parameterization. It is also clear that $\mathbf{T}(I)$ equals one if and only if $I = \mathrm{M}$ is the maximal ideal.

Suppose, on the other hand, the ideal defines a positive dimensional variety. We wish to find a parameterized holomorphic curve in this variety. To do so, we can first consider any one-dimensional subvariety of it. Then we select an irreducible one-dimensional branch of this subvariety. Such branches are actually the images of parameterized holomorphic curves. The (special case of the) normalization theorem for one-dimensional varieties implies that any (irreducible) one-dimensional branch is the image of an open subset of $\mathbb{C}$ under a parameterized holomorphic curve. It is therefore a consequence of the normalization theorem [Wh] that $\mathbf{T}(I) = \infty$ whenever $\mathbf{V}(I)$ is nontrivial. This also follows easily from Theorem 1.23. Later we will prove sharp inequalities relating $\mathbf{T}(I)$ to the other invariants.

Let us find $\mathbf{T}(I)$ for some examples in two variables.

Example 4
Put

$$I = \left(z^a, w^b\right). \tag{120}$$

Then $\mathbf{T}(I)$ is the maximum of a and b. The simple verification is left to the reader.

Next put

$$I = \left(z^2 - w^5, z^3 w\right). \tag{121}$$

It then follows that

$$\mathbf{T}(I) = \frac{17}{2} \tag{122}$$

because the only meaningful candidates for curve of largest order of contact are

$$\left(t^5, t^2\right) \quad v = 17$$
$$(t, 0) \quad v = 2$$
$$(0, t) \quad v = 5 \tag{123}$$

and the first curve has multiplicity two. We offer a bit more detail. One first considers curves defined by $z(t) = \left(t^k u(t), t^j v(t)\right)$ for integer exponents and units u, v. Then generically the lowest possible order of vanishing will be one of the numbers $2k, 5j, 3k + j$. Since $2k < 3k + j$, only the first two orders of vanishing are relevant. The interesting case is when $2k = 5j$, so cancellation can occur, and the order of vanishing depends on the units. After simple power series computations, we discover the candidate $\left(t^5, t^2\right)$. There are many other curves of multiplicity two also for which $v(z^* g) = 17$ for a generic element $g \in I$ and hence $\mathbf{T}(I, z) = 17/2$, but this is the maximum value. One also has to consider the curves $(t, 0)$ and $(0, t)$. In the end there are only finitely many rational numbers to compare. ▯

The following simple observation is useful.

REMARK 4 Since the definition of $\mathbf{T}(I, z)$ requires choosing the infimum over all elements of the ideal, the generic element f in an ideal I for which $\mathbf{T}(I)$ is finite satisfies

$$v\left(z^* f\right) = v(z)\,\mathbf{T}(I, z). \quad \blacksquare \tag{124}$$

2.3.2 The order of contact for ideals in other rings

The main application of the order of contact for an ideal in O is that it provides a mechanism for passing from orders of contact of real objects with complex varieties to the intersection numbers of Section 2.2. Thus it is now convenient to give the definition for more general rings.

We use the following notation. Let $\Lambda = \Lambda_p$ denote any of the following rings at p in $\mathbb{C}^n$:

$$
\left\{
\begin{array}{lll}
O = \mathbb{C}\{z - p\} & = & \text{convergent power series} \\
\mathbb{C}\{z - p, \overline{z - p}\} & = & \text{conv. pwr. series in } z - p, \overline{z - p} \\
\mathbb{C}[[z - p]] & = & \text{formal pwr. series} \\
\mathbb{C}[[z - p, \overline{z - p}]] & = & \text{formal pwr. series in } z - p, \overline{z - p} \\
\mathbb{C}_p^\infty & = & \text{germs of smooth fns.}
\end{array}
\right.
\tag{125}
$$

DEFINITION 9 *Suppose J is a proper ideal in Λ_p, and z is a parameterized holomorphic curve. Put*

$$
\mathbf{T}(J, z) = \inf_{g \in J} \frac{v(z^* g)}{v(z)} .
\tag{126}
$$

Then the order of contact of J is the positive real number defined by

$$
\mathbf{T}(J) = \sup_z \mathbf{T}(J, z) = \sup_z \inf_{g \in J} \frac{v_p(z^* g)}{v_p(z)} .
\tag{127}
$$

Note that when $\mathbf{T}(J, z)$ is finite, it makes sense to compute the orders of vanishing by substitution of a sufficiently high polynomial truncation of a power series. Thus there is no convergence difficulty.

The next lemma, albeit almost trivial, gives a glimpse of the relationship between the different versions of this invariant when several rings are involved.

LEMMA 5
Suppose $I = (f_1, \ldots, f_N)$ is an ideal in O. Write $J = (\|f\|^2)$ for the ideal in $\mathbb{C}\{z - p, \overline{z - p}\}$ generated by the norm-squared of the generators of I. Then

$$
\mathbf{T}(J) = 2\,\mathbf{T}(I).
\tag{128}
$$

PROOF For any z it is clear that

$$
v\left(z^* \|f\|^2\right) = v\left(\|z^* f\|^2\right) = 2 \min \left(v\left(z^*(f_j)\right)\right).
\tag{129}
$$

Since each f_j lies in the ideal, the right side of (129) is at least as large as

$$
2 \inf \left(v\left(z^* g\right)\right).
\tag{130}
$$

On the other hand, since the order of vanishing of a sum is at least the order of vanishing of each term, and the order of vanishing of a product equals the sum of the orders of vanishing, we see that, for $g = \sum a_j f_j$,

$$
v\left(z^* g\right) \geq \min v\left(z^*(a_j)\right) v\left(z^*(f_j)\right) \geq \min v\left(z^*(f_j)\right).
\tag{131}
$$

The infimum (129) over all such g therefore equals the right side of (131). Dividing by $v(z)$ and taking the supremum, the result follows. ∎

It is therefore quite easy to understand the order of contact for principal ideals generated by real-valued functions of the form $\| f \|^2$. Lemma 5 reduces this case to an understanding of the order of contact of an ideal of holomorphic functions. This technique works in complete generality, because of our decomposition (see Section 3.3.1) of real analytic, real-valued functions into sums and differences of squared norms.

2.3.3 Relationship between intersection numbers and orders of contact

The order of contact is distinct from the intersection numbers we have seen so far. For specific ideals it is often not easy to uncover the relationship between these numbers. On the other hand, appropriate inequalities show that these numbers are in some sense equivalent measurements. Since the order of contact is defined for ideals, we do not have to restrict consideration to the equidimensional case. Suppose that

$$f : \left(\mathbb{C}^n, 0 \right) \to \left(\mathbb{C}^N, 0 \right) \tag{132}$$

is the germ of a holomorphic mapping. We write $\mathbf{T}(f)$ for the order of contact of the ideal generated by the components. Our next goal is to establish the following sharp inequalities.

THEOREM 2
Suppose that

$$f : \left(\mathbb{C}^n, 0 \right) \to \left(\mathbb{C}^N, 0 \right) \tag{133}$$

is the germ of a holomorphic mapping and rank $df(0) = q$. Then the following (sharp) inequalities hold:

$$\mathbf{T}(f) \leq \mathbf{K}(f) \leq \mathbf{D}(f) \leq \mathbf{T}(f)^{n-q} \tag{134}$$

$$\mathbf{D}(f) \leq \binom{n - q + \mathbf{K}(f) - 1}{\mathbf{K}(f) - 1}. \tag{135}$$

The hardest of these inequalities is the statement that

$$\mathbf{D}(f) \leq \mathbf{T}(f)^{n-q}. \tag{136}$$

We give some examples where $q = 0$ to show that this estimate is sharp.

Example 5
Put $z = (a, b, c)$ and put $f(z) = \left(a^3 - b^3, b^3 - c^3, abc \right)$. Then

$$\mathbf{D}(f) = 27 = \mathbf{T}(f)^3. \tag{137}$$

An easy way to see this is as follows: By statement 7 of Theorem 1, and because the intersection number depends on the ideal and not on the choice of generators,

$$
\begin{aligned}
\mathbf{D}(f) &= \mathbf{D}\left(a^3 - b^3, b^3 - c^3, abc\right) \\
&= \mathbf{D}\left(a^3 - b^3, b^3 - c^3, a\right) + \mathbf{D}\left(a^3 - b^3, b^3 - c^3, b\right) \\
&\quad + \mathbf{D}\left(a^3 - b^3, b^3 - c^3, c\right) \\
&= \mathbf{D}\left(b^3, c^3, a\right) + \mathbf{D}\left(a^3, c^3, b\right) + \mathbf{D}\left(a^3, b^3, c\right) \\
&= 9 + 9 + 9 = 27.
\end{aligned}
\tag{138}
$$

We also substitute $\mathbf{D}(a, b, c) = 1$ to find that $\mathbf{D}(f) = 27$. On the other hand, it is obvious because of homogeneity that $\mathbf{T}(f) = 3$. ◻

The next result generalizes Example 5.

PROPOSITION 8

Suppose that I is an ideal generated by homogeneous polynomials all of degree m. Then either

$$
\mathbf{T}(I) = \mathbf{D}(I) = \infty
\tag{139}
$$

or

$$
\begin{aligned}
\mathbf{T}(I) &= m \\
\mathbf{D}(I) &= m^n.
\end{aligned}
\tag{140}
$$

Suppose I is any ideal for which $\mathbf{T}(I) < \infty$. Then I can be generated by homogeneous polynomials if and only if

$$
\mathbf{D}(I) = \mathbf{T}(I)^n .
\tag{141}
$$

Thus the sharp bound occurs only in the homogeneous case.

We indicate the proof of Proposition 8 at the end of Section 3.6. Let us now give an example showing that the invariants $\mathbf{T}, \mathbf{K}$ are not upper semicontinuous functions of parameters. This prevents them from being universally applicable.

Example 6

We give several families of ideals that depend on parameters, and determine whether they satisfy upper and lower semicontinuity in those parameters. We write our variables as $z = (a, b, c)$. Put

$$I_t = \left(a^3 - tbc, b^2, c^2\right)$$
$$J_s = \left(a^2 - sb, b^2, c\right)$$
$$Y_\epsilon = \left(\epsilon a + a^2, b, c\right). \tag{142}$$

Then the following hold:

$$\mathbf{K}(I_t) = 6 \text{ for } t > 0$$
$$\mathbf{K}(I_o) = 5$$
$$\mathbf{T}(J_s) = 4 \text{ for } s > 0$$
$$\mathbf{T}(J_o) = 2$$
$$\mathbf{K}(Y_\epsilon) = \mathbf{T}(Y_\epsilon) = 1 \text{ for } \epsilon > 0$$
$$\mathbf{K}(Y_o) = \mathbf{T}(Y_o) = 2. \tag{143}$$

These examples reveal that neither invariant is upper semicontinuous from either side. The failure of lower semicontinuity is to be expected, and has no particular consequences. The failure of upper semicontinuity has some interesting consequences. In particular we will construct a domain in Chapter 6 from the ideals J_s where the failure of semicontinuity has a consequence in the theory of subelliptic estimates. The reader can verify the values obtained in (143); we omit the details. It is useful in computing $\mathbf{K}$ or when listing a basis for O/I to remember that $\det(df) \not\equiv 0 \mod (f)$ if $I = (f)$ is generated by a regular sequence. This was Corollary 2. It is furthermore true in this context that $h \det(df) \equiv 0 \mod (f) \; \forall h \in \mathrm{M}$. See [EL]. $\quad \square$

2.3.4 More general intersection indices

The theory of intersections of varieties is highly developed [Fu]. We discuss just enough of this theory to enable us to relate the codimension of an ideal to its order of contact. We need the formula in Theorem 3 below, whose statement requires a more general notion of intersection index than we have given so far. Suppose V, W are irreducible complex analytic varieties, and they are in "general position." This means that their intersection is either empty or that all the irreducible components Z_i of their intersection are of the same dimension, and

$$\mathrm{cod}(Z_i) = \mathrm{cod}(V) + \mathrm{cod}(W). \tag{144}$$

Here $\mathrm{cod}(V)$ denotes the codimension of the variety in $\mathbb{C}^n$, namely $n - \dim(V)$. In this case it is possible to assign integer multiplicities to each component of the intersection. One writes

$$V \cap W \simeq \sum_{i=1}^{s} m_i Z_i \tag{145}$$

and calls the right side of (145) the intersection product of V and W. In general it requires complicated constructions from homological algebra to define the multiplicities m_i. In our applications, the varieties are defined by regular sequences, and the multiplicities are lengths of certain modules. In these cases, the theory parallels the case where we have defined the multiplicity of an isolated zero of a holomorphic mapping.

Suppose that (f) is a regular sequence in $_nO$ of order n. Then the variety $\mathbf{V}(f_1, \ldots, f_{n-1})$ defined by the first $n - 1$ functions is one-dimensional by Proposition 7. We can decompose this variety into a sum of irreducible branches, together with multiplicities that describe the number of times each branch is defined. For the moment we assume that we are able to define these multiplicities. We return to the definition later. Calling the irreducible branches

$$Z_k \quad k = 1, \ldots, s \tag{146}$$

we can write (formally)

$$V = \mathbf{V}\left(f_1, \ldots, f_{n-1}\right) \simeq \sum_{k=1}^{s} m_k Z_k \tag{147}$$

for positive integers m_k. For each Z_k, we can find a curve of minimal order that parameterizes it. This is a consequence of the normalization theorem. Let us call z_k the minimal parameterization of Z_k. It is possible to compute the codimension of the ideal (f) in terms of the Z_k and the m_k. The following formula gives the precise relationship:

THEOREM 3
For any function f for which $\mathbf{D}\left(f_1, \ldots, f_{n-1}, f\right)$ is finite,

$$\mathbf{D}\left(f_1, \ldots, f_{n-1}, f\right) = \sum_{j=1}^{s} m_j v\left(z_j^* f\right). \tag{148}$$

PROOF For a complete proof, see [Sfv, pp. 190–193]. Here we give an informal discussion.

We begin this discussion by recalling what it means to localize at a prime ideal. Start with the local ring $_nO_o = O$. If P is a prime ideal in O, we form the ring A_P as follows. We consider equivalence classes of fractions g/h, where both the numerator and denominator are in O, but the denominator does not lie in P. Two fractions g/h and g'/h' are equivalent if $gh' - hg'$ is a divisor of zero in O/P. One defines addition and multiplication in the usual way that one adds and multiplies fractions. The resulting A_P is a Noetherian local ring, where we can divide by elements not in the given prime ideal.

Recall next that the length of a module N is defined as follows. Write

$$N = N_0 \supset N_1 \supset \cdots \supset N_m = 0 \tag{149}$$

for submodules N_k such that $N_k \neq N_{k+1}$ and such that N_k/N_{k+1} is simple. From the Jordan–Holder theorem, the number m in any chain such as (149) is independent of the chain. This integer is the length of N. The notion of length enjoys properties very similar to those for dimension of a vector space; in particular if the ground ring is a field, the two notions are the same.

Suppose now that $(f_1, \ldots, f_{n-1})$ is a regular sequence of order $n - 1$. As above, these functions define a one-dimensional variety, whose irreducible branches Z_k can also be defined by prime ideals P_k. The number m_k is then defined to be the length of the quotient module $A_{P_k}/(f_1, \ldots, f_{n-1})$. The length represents in some sense the number of times that the equations $(f_1, \ldots, f_{n-1})$ define the same set as does the prime ideal P_k. This is analogous to the statement that a regular sequence (f) of order n defines the origin $\mathbf{D}(f) = \dim(O/(f))$ times, whereas the maximal ideal defines it but once. Putting all this intuition together suggests but does not prove (148). See [Fu,Sfv] for details. ∎

2.3.5 An explicit computation

In order to get a feeling for these ideas, we apply the formula of Theorem 3 to a concrete problem. We compute the same number in several ways. The purpose of these computations is not to derive the multiplicities as efficiently as possible, but rather to shed light on the formula. We write the variables in $\mathbb{C}^4$ as (a, b, c, d) so as to avoid having too many indices. Consider the ideal

$$I = (a^3 - bcd, b^3 - acd, c^3 - abd). \tag{150}$$

The three functions define a one-dimensional variety in $\mathbb{C}^4$. Let us find all its branches through the origin and their multiplicities. It is intuitively clear, and easy to check, that any one-dimensional branch is a complex line. Call such a line (At, Bt, Ct, Dt) and note that if D is not zero, then we may assume without loss of generality that it is unity. It does not vanish, because

$$D = 0 \Rightarrow A = B = C = 0 \tag{151}$$

also. Substituting (At, Bt, Ct, t) into the defining equations gives the relations:

$$A^3 = BC$$
$$B^3 = AC$$
$$C^3 = AB. \tag{152}$$

One solution is $(0,0,0)$. This yields the line parameterized by $(0,0,0,t)$. Otherwise we find by multiplying the equations together that

$$1 = ABC = A^4 = B^4 = C^4. \tag{153}$$

There are sixteen solutions to these equations, corresponding to the choice of a pair of fourth roots of one. Let us number the corresponding lines:

$$Z_o = (0,0,0,t)$$
$$Z_1 = (t,t,t,t)$$
$$Z_j = (At, Bt, Ct, t) \quad j = 2, \ldots, 16. \tag{154}$$

We want to compute the multiplicities with which they are defined. One way to do this is to choose test functions $g(a,b,c,d)$ such that the three generators of I together with g define a regular sequence of order four, then compute the intersection number in two ways. Treating the multiplicities as variables, we can solve a system of equations for them. First choose $g(a,b,c,d) = d$. Then

$$\mathbf{D}(f_1, f_2, f_3, g) = \mathbf{D}\left(a^3, b^3, c^3, d\right) = 27 \tag{155}$$

where we have used properties 2 and 5 of Theorem 1. By our formula we also have

$$\mathbf{D}(f_1, f_2, f_3, g) = \sum_{j=0}^{16} m_j v\left(Z_j^* g\right). \tag{156}$$

We claim that

$$m_o = 11$$
$$m_j = 1 \quad j = 1, \ldots, 16. \tag{157}$$

One could obtain these values directly, but we derive them in a manner that fully illustrates the power of formula (148). These multiplicities are defined to be lengths of modules, but it is possible to compute them without using any of the ideas from Jordan–Holder decompositions of modules. The idea is to use enough test functions in the last slot to obtain a system of linear equations for the m_j. Knowing that the solutions are positive integers simplifies the work even more. First we choose the test function to be d. It is easy to compute that

$$\mathbf{D}\left(a^3 - bcd, b^3 - acd, c^3 - abd, d\right) = \mathbf{D}\left(a^3, b^3, c^3, d\right) = 3^3 = 27. \tag{158}$$

The first step follows because $\mathbf{D}(f)$ depends only on the ideal; the second step follows from the additivity (property 7 of Theorem 1). On the other hand, according to formula (148),

$$\mathbf{D}\left(a^3 - bcd, b^3 - acd, c^3 - abd, d\right) = \sum_{j=0}^{16} m_j v\left(z_j^* d\right) = \sum_{j=0}^{16} m_j \tag{159}$$

because the test function vanishes to first order when pulled back to any of the seventeen curves. Next we choose a test function that distinguishes the curve Z_o from the rest. An example is $a - d^2$. Using formula (148) again, together with the previous result, we obtain

$$\mathbf{D}\left(a^3 - bcd, b^3 - acd, c^3 - abd, a - d^2\right) = 2m_o + \sum_{j=1}^{16} m_j = 27 + m_o. \quad (160)$$

Therefore, if we could compute this intersection number directly, we could find m_o. To do so, we first substitute $a = d^2$. Then we observe that a local biholomorphic coordinate change enables us to replace the last slot by a linear function. We then reduce dimensions by one to get

$$\mathbf{D}\left(a^3 - bcd, b^3 - acd, c^3 - abd, a - d^2\right)$$
$$= \mathbf{D}\left(d^6 - bcd, b^3 - cd^3, c^3 - bd^3, a - d^2\right)$$
$$= \mathbf{D}\left(d^6 - bcd, b^3 - cd^3, c^3 - bd^3\right). \quad (161)$$

Now the first component factors, so we use additivity, obtaining

$$\mathbf{D}\left(a^3 - bcd, b^3 - acd, c^3 - abd, a - d^2\right)$$
$$= \mathbf{D}\left(d^6 - bcd, b^3 - cd^3, c^3 - bd^3\right)$$
$$= \mathbf{D}\left(d, b^3 - cd^3, c^3 - bd^3\right) + \mathbf{D}\left(d^5 - bc, b^3 - cd^3, c^3 - bd^3\right)$$
$$= \mathbf{D}\left(d, b^3, c^3\right) + \mathbf{D}\left(d^5 - bc, b^3 - cd^3, c^3 - bd^3\right). \quad (162)$$

Since the first term in the last row obviously equals 9, we are reduced to computing

$$\mathbf{D}\left(d^5 - bc, b^3 - cd^3, c^3 - bd^3\right). \quad (163)$$

We apply the same technique to this simpler situation. The variety defined by the first two functions has five branches. These are the curves:

$$(b, c, d) = (0, t, 0) \qquad \text{or}$$
$$(b, c, d) = \left(\delta^3 t^2, \delta t^3, t\right) \quad \text{where} \quad \delta^4 = 1. \quad (164)$$

Call the multiplicities of these curves n_o, n_1, n_2, n_3, n_4. By symmetry considerations the last four of these are equal, but we do not use this fact. Using the test function c gives

$$15 = \mathbf{D}\left(d^5 - bc, b^3 - cd^3, c\right) = n_0 + 3\sum_{k=1}^{4} n_k. \quad (165)$$

The only solution in positive integers to this equation is

$$n_o = 3$$

$$n_j = 1 \qquad j = 1, \ldots, 4. \tag{166}$$

Using the formula (148) for the last time, and then inserting these values gives

$$\mathbf{D}\left(d^5 - bc, b^3 - cd^3, c^3 - bd^3\right) = 3n_0 + 5\sum_{j=1}^{4} n_j = 29. \tag{167}$$

Now we are done. Putting together (159), (162), and (167) we see that

$$m_o + 27 = 9 + 29$$

$$m_0 = 11. \tag{168}$$

Since the other multiplicities are positive integers, they are all equal to one. ∎

REMARK 5   In this explicit computation, the simplest way to verify that $m_0 = 11$ is to compute the module $A_P/(f_1, f_2, f_3)$ directly. The prime ideal that defines the curve Z_0 is (a, b, c). The localization process guarantees that we may divide by the fourth coordinate d. It is then possible to work in three variables. More precisely, in the localization we may divide, so we may divide each of the generators by d^3. Replace each of the first three variables by their quotients upon division by d, but keep the same notation. The resulting generators become $\left(a^3 - bc, b^3 - ac, c^3 - ab\right)$; essentially we may simply set $d = 1$. It is then easy to see that $m_0 = \mathbf{D}\left(ab - c^3, ac - b^3, bc - a^3\right)$. Notice that of the 17 curves defined by the original sets of equations, only the curve Z_0 passes through the origin in this affine subspace. This is why it is sufficient to compute $\mathbf{D}\left(ab - c^3, ac - b^3, bc - a^3\right)$. This can be done in any of the ways we have discussed. Using Definition 4 (the dimension of the quotient space) is easiest, as one can list the basis elements of the quotient:

$$1, a, b, c, a^2, b^2, c^2, ab, ac, bc, abc. ∎ \tag{169}$$

More generally, suppose that $(f_1, \ldots, f_{n-1})$ is a regular sequence, and $\mathbf{V}(f_1, \ldots, f_n) \simeq \sum m_j Z_j$. In case Z_1 defines a smooth curve, we can choose coordinates so that this curve is given by the prime ideal $(z_1, \ldots, z_{n-1})$. Then the multiplicity m_1 has a precise interpretation as the number of times that $(f_1, \ldots, f_{n-1})$ defines $\mathbf{V}(z_1, \ldots, z_{n-1})$. To see this, observe that localization at this prime ideal amounts to treating z_n as a parameter. Let us write f_j^* for the function obtained from f_j after localizing. Computing the length of the module is then the same as computing the intersection number $\mathbf{D}(f_1^*, \ldots, f_{n-1}^*)$ in the space of $n - 1$ variables.

2.3.6 Inequalities among the invariants

Now that the notion of regular sequence is established, and we have seen how to apply formula (148), it becomes possible to prove Theorem 2 and Proposition 8. The hard part of the proof of Theorem 2 is the statement that

$$\mathbf{D}(I) \leq \mathbf{T}(I)^{n-q}. \tag{170}$$

Let us dispense with the other inequalities first.

PROOF OF THEOREM 2 Note first that all three invariants are inclusion reversing: if $I \subset J$, then $W(J) \subset W(I)$. Here W is any of $\mathbf{T}, \mathbf{D}, \mathbf{K}$. Since $\mathbf{T}\left(\mathrm{M}^k\right) = k$, the inequality $\mathbf{T}(I) \leq \mathbf{K}(I)$ follows immediately from the inclusion $\mathrm{M}^k \subset I$. The inequality $\mathbf{K}(I) \leq \mathbf{D}(I)$ follows from formula 7 of Theorem 1 as follows. Suppose that $\mathrm{M}^k \subset I$ but $\mathrm{M}^{k-1} \not\subset I$. Then there is some homogeneous monomial of degree $k-1$ that is not in I. It is an easy exercise to verify that a monomial of degree $k-1$ is a linear combination of $(k-1)$st powers of linear functions. Thus there must be some linear function h for which the powers

$$1, h, h^2, \ldots, h^{k-1} \tag{171}$$

are linearly independent elements of O/I. This implies, though, that $\mathbf{D}(I) \geq k$. The inequality that

$$\mathbf{D}(I) \leq \binom{n-1-q+\mathbf{K}(f)}{\mathbf{K}(f)-1} \tag{172}$$

follows from the inclusion reversing property of $\mathbf{D}$ together with an explicit count of $\dim\left(\dot{O}/\mathrm{M}^k\right)$. The combinatorially inclined reader can perform this count without difficulty. ∎

We state the remaining inequality as a separate theorem.

THEOREM 4
Suppose I is an ideal in $_nO$ and I contains q independent linear functions. Then the following sharp estimate holds:

$$\mathbf{T}(I) \leq \mathbf{D}(I) \leq (\mathbf{T}(I))^{n-q}. \tag{173}$$

PROOF The result is trivial if the variety of the ideal is positive dimensional, as all the numbers are then infinite. If I is the whole ring, then all the numbers vanish, and the result holds using the convention that $0^0 = 0$. The important case is when I defines a variety consisting of just one point. We verified the first inequality just after the statement of (170). To verify the second, we proceed as follows. By the Nullstellensatz, we can find a regular sequence of order n in I, for example by choosing sufficiently high powers of the coordinate functions. Consider such a regular sequence $(f) = (f_1, \ldots, f_n)$ of elements in I of smallest possible intersection number. This will be true for a generic choice, and it will include q independent linear functions. We may suppose that these are $(f) = (f_1, \ldots, f_q)$. By the inclusion reversing property, we have

$$\mathbf{D}(I) \leq \mathbf{D}(f). \tag{174}$$

It is then sufficient to prove that

$$\mathbf{D}(f) \leq (\mathbf{T}(I))^{n-q}. \tag{175}$$

Select the first $n-1$ of these functions; they form a regular sequence of order $n-1$ and thus define a one-dimensional variety. We write

$$V = \mathbf{V}(f_1, \ldots, f_{n-1}) \simeq \sum_{k=1}^{s} m_{1k} Z_{1k} \tag{176}$$

for its decomposition into irreducible branches Z_{1k}. Let z_{1k} denote the parameterized holomorphic curve of minimal order that normalizes Z_{1k}. For a generic linear function g_1, we have

$$\mathbf{D}(f_1, \ldots, f_{n-1}, g_1) = \sum_{j=1}^{s_1} m_{1j} v\left(z_{1j}^* g_1\right)$$

$$= \sum_{j=1}^{s_1} m_{1j} v\left(z_{1j}\right). \tag{177}$$

Now consider

$$\mathbf{D}(f) = \mathbf{D}(f_1, \ldots, f_n). \tag{178}$$

By the formula (148) and the definition of the order of contact,

$$\mathbf{D}(f_1, \ldots, f_{n-1}, f_n) = \sum_{j=1}^{s_1} m_{1j} v\left(z_{1j}^* f_n\right)$$

$$\leq \sum_{j=1}^{s_1} m_{1j} v\left(z_{1j}\right) \mathbf{T}(I)$$

$$= \mathbf{D}(f_1, \ldots, f_{n-1}, g_1) \mathbf{T}(I). \tag{179}$$

Now, since g_1 is a nontrivial linear function, we can compute the intersection number

$$\mathbf{D}(g_1, f_1, f_2, \ldots, f_{n-1}) \tag{180}$$

by the same method. Again there are finitely many branches, and we know that

$$v\left(z_{2j}^{*}\left(f_{n-1}\right)\right) \leq v\left(z_{2j}\right) \mathbf{T}\left(I\right) \tag{181}$$

because the function f_{n-1} lies in I. We proceed in this fashion until we obtain

$$\mathbf{D}\left(g_1, g_2, \ldots, g_{n-q}, f_1, \ldots, f_q\right) \tag{182}$$

which equals unity, because the ideal $\left(g_1, \ldots, g_{n-q}, f_1, \ldots, f_q\right)$ is generated by n independent linear functions. Putting the resulting inequalities together shows that a factor of $\mathbf{T}\left(I\right)^{n-q}$ results. This gives the desired inequality. ∎

From the proof one can determine when equality holds. Suppose for simplicity that $q = 0$. In order that equality occurs, the order of vanishing of the last function, when restricted to each curve, must be maximal. This means that every one of the curves that arises in any of the decompositions gives the same number for the order of vanishing of the pullback. Using induction on the dimension one verifies that the ideal must be homogeneous. On the other hand, it is easy to verify that, for ideals generated by homogenous polynomials of degree m, either $\mathbf{D}\left(I\right) = \mathbf{T}\left(I\right) = \infty$ or $\mathbf{T}\left(I\right) = m$, $\mathbf{D}\left(I\right) = m^{n}$. (The second statement is essentially the Bezout theorem.) These statements imply Proposition 8. ∎

2.4 Higher order invariants

2.4.1 Definition of the invariants

The proof of Theorem 4 shows the importance of knowing how a variety sits relative to linear subspaces. Together with the local parameterization theorem, it motivates the following definitions of higher order invariants.

DEFINITION 10 *Suppose that* $f : (\mathbb{C}^{\mathbf{n}}, 0) \rightarrow (\mathbb{C}^{\mathbf{N}}, 0)$ *is the germ of a holomorphic mapping. For each integer* $q : 1 \leq q \leq n$ *we define invariants of the ideal* (f) *as follows:*

$$\mathbf{D_q}\left(f\right) = \inf \mathbf{D}\left(f_1, \ldots f_N, l_1, \ldots, l_{q-1}\right)$$
$$\mathbf{T_q}\left(f\right) = \inf \mathbf{T}\left(f_1, \ldots f_N, l_1, \ldots, l_{q-1}\right) \tag{183}$$

where the infima are taken over all choices of $q-1$ *linear functions* $\{l_1, \ldots, l_{q-1}\}$.

This amounts to restricting the functions to all possible linear subspaces of codimension $q-1$, applying the previous definitions, and then taking the minimal value over all such choices of subspaces. Now that we have this notation, we will no longer use a subscript on $\mathbf{D}$ to denote a base point. Recall that we had needed such notation in Section 1.6.

It is now elementary to verify that the numbers $\mathbf{D_q}$ and $\mathbf{T_q}$ satisfy the following properties:

$$\mathbf{T_q}(f) \leq \mathbf{D_q}(f) \leq \mathbf{T_q}(f)^{n-q+1}$$

$$\mathbf{T_1}(f) = \mathbf{T}(f)$$

$$\mathbf{D_1}(f) = \mathbf{D}(f)$$

$$\mathbf{T_n}(f) = \mathbf{D_n}(f) = \min\{v(g) : g \in (f)\}$$

$$\mathbf{D_q}(f) < \infty \Leftrightarrow \dim \mathbf{V}(f) < q. \tag{184}$$

Most of these properties follow immediately from the special case when q equals unity by working in a complex vector subspace of dimension $n-q+1$. For example, $\mathbf{D_n}(f) = \min\{v(g) : g \in (f)\}$ follows because the one-dimensional intersection number equals the order of the zero. The fact that $\mathbf{D_n}(f) = \mathbf{T_n}(f)$ follows from the first line of (184). We summarize all these conclusions in Theorem 5. This result shows that we have completed the aims of this chapter.

THEOREM 5
The intersection numbers $\mathbf{D_q}$ and the orders of contact $\mathbf{T_q}$ assign to an ideal (f) in O numbers $\#_q(f)$ (or plus infinity) satisfying the following properties:

$$\#_1(f) \geq \#_2(f) \geq \cdots \geq \#_{n-1}(f) \geq \#_n(f)$$

$$\#_q(f) < \infty \Leftrightarrow \dim(\mathbf{V}(f)) < q$$

$$\#_q(f) = 1 \Leftrightarrow \operatorname{rank}(df(0)) > n - q. \tag{185}$$

In addition, the sharp inequalities

$$\mathbf{T_q}(f) \leq \mathbf{D_q}(f) \leq \mathbf{T_q}(f)^{n-q+1} \tag{186}$$

all hold. If the ideal (f) contains m independent linear functions, and $m \leq n - q + 1$, then the exponent on the right in (186) can be improved to $n - q + 1 - m$. The number $\mathbf{T_n}(f) = \mathbf{D_n}(f)$ equals the minimum order of vanishing of any function in the ideal. Finally, the function $(f_\epsilon) \to \mathbf{D_q}(f_\epsilon)$ is upper semicontinuous as a function of the parameter ϵ.

It is worthwhile to give an example where we compute the higher order invariants.

Example 7
Consider the ideal in $_3O$ generated by

$$(f) = \left(\left(z_1^3 + z_2^3 + z_3^3\right)^2, z_3^7 \right).$$

The variety of the ideal is obviously one-dimensional. Thus we have $D_1(f) = T_1(f) = \infty$. Following (145) we write $V(f) \simeq 14Z_1 + 14Z_2 + 14Z_3$, where each branch is a line given by

$$z_1 = -\omega z_2 \quad \text{and} \quad z_3 = 0$$
$$\omega^3 = 1 \tag{187}$$

and the lengths are easy to find. The first equation is defined twice and the second is defined seven times. Next we claim that

$$D_2(f) = 42, \quad D_1(f) = 6$$
$$T_2(f) = 7, \quad T_1(f) = 6. \tag{188}$$

To verify these, we adjoin linear functions as in the definition and take the infimum. Thus, for example,

$$T_2(f) = T_1\left(z_1, \left(z_1^3 + z_2^3 + z_3^3\right)^2, z_3^7\right)$$
$$= T_1\left(z_1, \left(z_2^3 + z_3^3\right)^2, z_3^7\right) = 7 \tag{189}$$

where a curve of highest contact is $(0, t, -t)$.

If we consider the first generator f_1 alone, we then have $T_1(f_1) = T_2(f_1) = \infty$ and $T_3(f_1) = 6$. Note that $V(f_1)$ is two-dimensional and is defined by the square of an irreducible Weierstrass polynomial. Pieces of it can be parameterized by formulas such as

$$z_1(s, t) = t$$
$$z_2(s, t) = st$$
$$z_3(s, t) = -t\left(1 + s^3\right)^{1/3} \tag{190}$$

but there does not exist a parameterization of all of $V(f_1)$. The given formula misses, for example, the three lines given by $(0, \zeta, -\omega\zeta)$ where $\omega^3 = 1$ and $\zeta \in \mathbb{C}$. For other attempts at parameterization, a similar phenomenon will invariably result.　□

3

Geometry of Real Hypersurfaces

3.1 CR geometry

3.1.1 The Levi form

Suppose that M is a real submanifold of $\mathbb{C}^n$. The tangent spaces of M then inherit a certain amount of complex structure from the ambient space $\mathbb{C}^n$. The resulting structure is known as a CR structure, although there is some debate about whether CR stands for "Cauchy–Riemann" or "complex-real." We review the appropriate ideas.

Let $T\mathbb{C}^n$ denote the tangent bundle of $\mathbb{C}^n$ and $\mathbb{C}T\mathbb{C}^n = T\mathbb{C}^n \otimes \mathbb{C}$ its complexification. Complexification of a bundle has the effect of allowing local sections to have complex coefficients. A complex vector field X on $\mathbb{C}^n$ is a functional combination of the first-order differential operators defined in Chapter 1:

$$X = \sum_{j=1}^{n} a_j \frac{\partial}{\partial z^j} + \sum_{j=1}^{n} b_k \frac{\partial}{\partial \bar{z}^k} . \tag{1}$$

The coefficient functions are smooth and complex-valued. This formula shows that it is natural to write $\mathbb{C}T\mathbb{C}^n$ as a direct summand:

$$\mathbb{C}T\mathbb{C}^n = T^{1,0}\mathbb{C}^n + T^{0,1}\mathbb{C}^n . \tag{2}$$

where $T^{1,0}\mathbb{C}^n$ denotes the subbundle whose sections are combinations of the $\partial/\partial z^j$, and $T^{0,1}\mathbb{C}^n$ denotes its complex conjugate bundle. More generally we write $\mathbb{C}TM = TM \otimes \mathbb{C}$ for the complexification of the tangent bundle of any smooth or complex manifold M. Then (2) applies if we replace $\mathbb{C}^n$ by any complex manifold W.

Let us return to the real submanifold M. One defines $T^{1,0}M$, the bundle of $(1,0)$ vectors on M, to be the intersection

$$T^{1,0}M = \mathbb{C}TM \cap T^{1,0}\mathbb{C}^n . \tag{3}$$

There is a similar definition of its conjugate $T^{0,1}M$. Recall [La] that the Lie bracket of two vector fields X and Y is the vector field $[X, Y]$ defined by $[X, Y](f) = (XY - YX)(f)$. Each of the bundles $T^{1,0}M$ and $T^{0,1}M$ is closed under the Lie bracket operation. To see this, note that the bracket of two vector fields that are each sections of $T^{0,1}\mathbb{C}^n$ is also such a section, and if each vector field is in addition tangent to M, the bracket is also tangent to M.

Our main interest is in the boundaries of domains in $\mathbb{C}^n$. If such a boundary M is a smooth manifold, then it is a real hypersurface. This means that, considered as a submanifold, it has real codimension one. The following properties of the bundle of (1,0) vectors on M are the defining properties of the CR structure on a real hypersurface in any complex manifold:

1. $T^{1,0}M$ is integrable, i.e., it is closed under the Lie bracket operation.

2. $T^{1,0}M \cap T^{0,1}M = \{0\}$.

3. $HM = T^{1,0}M + T^{0,1}M$ is a bundle of codimension one in $\mathbb{C}TM$. This phrase means that, at each point p, the complex vector space H_pM has codimension one in $\mathbb{C}T_pM$.

DEFINITION 1 (CR manifolds, CR functions, and CR mappings). *A CR manifold is a real manifold M whose complexified tangent bundle contains an integrable subbundle $T^{1,0}M$ that satisfies properties 1 and 2 above. The codimension of the CR structure is the codimension of HM. In case the codimension of a CR structure is unity, the CR manifold is of "hypersurface type." A CR function f is a continuous function such that $\overline{L}f = 0$ (in the distribution sense) for all smooth local sections $\overline{L}$ of $T^{0,1}(M)$. A CR mapping is a smooth mapping between CR manifolds M and M' whose differential maps $T^{1,0}M$ into $T^{1,0}M'$.*

The notion of CR manifold is an abstraction of the notion of real submanifold of $\mathbb{C}^n$. Not all CR manifolds can be imbedded in some $\mathbb{C}^n$. See [Ro2] and the references therein for recent results. The CR functions are generalizations of the (restrictions of) holomorphic functions. The simplest example of a CR function on a real submanifold of $\mathbb{C}^n$ that is not the restriction of a holomorphic function arises from considering a codimension-one real hyperplane M in $\mathbb{C}^n$. Suppose without loss of generality that this hyperplane is defined by the equation $\mathrm{Re}\,(z^n) = 0$. Then a basis for the sections of the bundle $T^{0,1}M$ are precisely the vector fields $\partial/\partial\bar{z}^j$, $j = 1, \ldots, n-1$, and consequently (the restriction to M of) any function of $z^n, \bar{z}^n$ is a CR function. Such a function cannot generally be extended to be holomorphic. The literature on the extension of CR functions to one or both sides of a hypersurface is extensive. See the book [Bg] and the references in the survey article [Fo3].

Associated with a real hypersurface (or with a CR structure of CR codimension one) is the Levi form λ; this is an Hermitian form on $T^{1,0}M$ that measures how far the bundle HM deviates from being integrable. Since HM has codimension one, it is the bundle annihilated by a nonvanishing one-form η. Such a one-form η is defined only up to a multiple, so the Levi form will be defined only up to a multiple. We choose a defining one-form η and suppose without loss of generality that it is purely imaginary. Let $\langle , \rangle$ denote the contraction of one-forms and vector fields, and let $[,]$ denote the Lie bracket of two vector fields. Suppose that L is a (1,0) vector field on M and $\overline{K}$ is a (0,1) vector field on M.

DEFINITION 2   *The Levi form is the Hermitian form on $T^{1,0}M$ defined by*

$$\lambda\left(L, \overline{K}\right) = \left\langle \eta, \left[L, \overline{K}\right]\right\rangle. \tag{4}$$

DEFINITION 3   *The real hypersurface M is pseudoconvex if λ is semidefinite, and is strongly pseudoconvex if λ is definite.*

Pseudoconvexity is the complex analogue of convexity. A domain is pseudoconvex if the function

$$z \to -\log\left(\text{dist}\left(z, b\Omega\right)\right) \tag{5}$$

is plurisubharmonic and continuous. See [Hm,Kr,Ra] for complete discussions of the various equivalent ways to define pseudoconvexity for a domain and the general definition of plurisubharmonic function. A smooth function f is plurisubharmonic if its complex Hessian $\partial\overline{\partial}f$ is positive semidefinite. We note the distinction between plurisubharmonicity of a defining function for a hypersurface and pseudoconvexity of the hypersurface. Pseudoconvexity requires semidefiniteness of the complex Hessian on a subspace of codimension one while plurisubharmonicity requires it on the full space. See also the discussion following (10).

The remarkable thing about the Levi form is that it affords a manner to decide whether a domain with smooth boundary is pseudoconvex. It is not difficult to show that a domain with smooth boundary is pseudoconvex in the sense (5) precisely when its boundary is pseudoconvex in the sense of Definition 3. It is fairly easy to prove that a domain of holomorphy is a pseudoconvex domain. [See Hm, Kr, Ra] for details on these matters. The converse question is the Levi problem; its solution is a highly nontrivial result. It states that the pseudoconvex domains are precisely the domains of holomorphy. Hence a domain with smooth boundary is a domain of holomorphy precisely when its boundary is pseudoconvex in the sense of Definition 3. A domain with smooth boundary is called strongly pseudoconvex if its boundary is strongly pseudoconvex in the sense of Definition 3.

In case M bounds a bounded domain in $\mathbb{C}^n$, there must always be at least one point where the Levi form is definite, namely a point on M at greatest distance from the origin. This is because the Levi form on a sphere is definite, and the sphere osculates M to order two at such a point. It is natural in this case to choose the sign so that the Levi form is positive definite there. It follows by continuity that there is an open neighborhood of this point where the Levi form remains positive definite.

There are several methods for computing the Levi form. Recall [La] that the Cartan formula for the exterior derivative of a one-form tells us that

$$\langle d\eta, L \wedge \overline{K} \rangle = L \langle \eta, \overline{K} \rangle - \overline{K} \langle \eta, L \rangle - \langle \eta, [L, \overline{K}] \rangle. \tag{6}$$

In case L, K are local sections of $T^{1,0}M$, their contractions with η vanish, and the Cartan formula implies that

$$\lambda(L, \overline{K}) = \langle -d\eta, L \wedge \overline{K} \rangle. \tag{7}$$

A second formula arises from choosing a local defining equation for M. We give now the definition of local defining function for a smoothly bounded domain in order to fix terminology.

Suppose that Ω is a domain in $\mathbb{C}^n$ and the boundary of Ω is a smooth manifold M. Then a smooth real-valued function r is a local defining function for Ω if M is locally given by the zero set of r. More precisely, suppose that p is a point in M. Then r is called a local defining function near p if there is a neighborhood U_p such that

$$r : U_p \to \mathbf{R} \text{ is } C^\infty$$

$$r < 0 \text{ on } \Omega \cap U_p$$

$$r = 0 \text{ on } M \cap U_p$$

$$r > 0 \text{ on } \Omega^c \cap U_p$$

$$dr \neq 0 \text{ on } U_p. \tag{8}$$

In this situation we always choose

$$\eta = \frac{1}{2} \left(\partial - \overline{\partial} \right) r \tag{9}$$

and, since $\lambda = -d\eta$, we see that

$$\lambda = \partial \overline{\partial} r. \tag{10}$$

Therefore the Levi form is the complex Hessian $(r_{i\overline{j}})$ of the defining function, restricted to the space of $(1,0)$ vectors tangent to the hypersurface. It is a standard fact that the Levi form is the biholomorphically invariant part of the real Hessian of the defining function; one way to derive it is seek a biholomorphically invariant analogue of Euclidean convexity. See for example [Kr]. From the definition of plurisubharmonic it follows that the level set of a smooth plurisubharmonic function is pseudoconvex, although there are pseudoconvex domains that do not have plurisubharmonic defining functions. Any strongly pseudoconvex domain has a strongly plurisubharmonic defining function, that is, one whose complex Hessian is definite. In all cases, if we choose a new defining function by multiplying r by a nonvanishing factor, then the Levi form on the zero set of r is changed only by multiplication by that factor. This statement is a consequence of Definition 2, which is an intrinsic definition of the Levi form (up to a multiple).

There is another differential form that is useful, and that depends on another choice. Once we have selected η, we can choose a purely imaginary vector field T, tangent to M, for which

$$\langle \eta, T \rangle = 1. \tag{11}$$

The direction defined by T is commonly called the "bad direction"; the reason is that estimates for derivatives in this direction are weaker than for the other tangential directions (see [GS]). The differential form that arises in this regard is a Lie derivative. Recall that the Lie derivative operator is defined by

$$L_X(\omega) = i_X d\omega + d(i_X \omega). \tag{12}$$

Here X is a vector field, i_X denotes interior multiplication, and ω is a differential form. When taking iterated commutators of $(1,0)$ vectors and their conjugates (see 4.3), certain annoying terms arise whose description is simplified by the use of the following one-form:

$$\alpha = \alpha_{(T)} = -L_T \eta. \tag{13}$$

We have the following lemma about this differential form.

LEMMA 1
The differential form α can be defined also by the formula

$$\langle \alpha, X \rangle = \langle \eta, [T, X - \langle \eta, X \rangle T] \rangle. \tag{14}$$

PROOF The result follows from the Cartan formula, the definition of interior multiplication, and because $\langle \eta, T \rangle = 1$ is constant. In fact,

$$\begin{aligned}
\langle \alpha, X \rangle &= -\langle L_T \eta, X \rangle \\
&= -\langle i_T d\eta, X \rangle - \langle d i_T \eta, X \rangle \\
&= -\langle d\eta, T \wedge X \rangle \\
&= -T\langle \eta, X \rangle + X\langle \eta, T \rangle + \langle \eta, [T, X] \rangle \\
&= -T\langle \eta, X \rangle + \langle \eta, [T, X] \rangle \\
&= \langle \eta, [T, X] \rangle - \langle \eta, [T, \langle \eta, X \rangle T] \rangle. \tag{15}
\end{aligned}$$

Note that the last step is checked most easily by reading from the bottom up. ∎

It is convenient to choose a local basis for the sections of $\mathbb{C}TM$ in terms of a local defining function. We suppose that the origin lies in M and, after a linear change of coordinates,

$$r(z, \bar{z}) = 2 \operatorname{Re}(z^n) + \text{ higher order terms.} \tag{16}$$

Then we define the local vector fields

$$L_i, \overline{L_j}, T \tag{17}$$

by

$$L_i = \frac{\partial}{\partial z^i} - \frac{r_{z_i}}{r_{z_n}} \frac{\partial}{\partial z^n} \tag{18}$$

$$T = \frac{1}{r_{z_n}} \frac{\partial}{\partial z^n} - \frac{1}{r_{\bar{z}_n}} \frac{\partial}{\partial \bar{z}^n} \, . \tag{19}$$

It is useful to express our notions in terms of this local basis. To simplify notation, we use subscripts to denote partial derivatives of r.

PROPOSITION 1
Let M be a real hypersurface of $\mathbb{C}^n$. The vector fields $L_i, \overline{L_j}, T$ defined by (17), (18), (19) form a local basis for (sections of) $\mathbb{C}TM$. They satisfy the following commutation relations:

1. $[L_i, L_k] = 0.$
2. $\left[L_i, \overline{L_j}\right] = \lambda_{i\bar{j}} T.$
3. $[T, L_i] = \alpha_i T.$
4. *The Levi form with respect to this basis is*

$$\lambda_{i\bar{j}} = \frac{r_{i\bar{j}} r_n r_{\overline{n}} - r_{i\overline{n}} r_n r_{\bar{j}} - r_{n\bar{j}} r_i r_{\overline{n}} + r_{n\overline{n}} r_i r_{\bar{j}}}{|r_n|^2} \tag{20}$$

   where the subscripts denote partial derivatives.
5. *The one-form α satisfies*

$$\alpha_i = \langle \alpha_{(T)}, L_i \rangle = \frac{r_n r_{i\overline{n}} - r_i r_{n\overline{n}}}{|r_n|^2} \, .$$

PROOF These are routine computations left to the reader. ∎

The Levi form is fundamental in studying the geometry of hypersurfaces. We are particularly interested in what happens when the Levi form is degenerate. It is evident that we must take more derivatives to get equivalent information. In what fashion should one take higher derivatives? Note the terrifying appearance of the formula (20) for the Levi form. One could imagine that methods applicable to higher derivatives will result in even more gruesome formulas, but they will not, as our more algebraic-geometric approach is rather elegant. Instead of obtaining information by iterated commutators, or derivatives of the Levi form itself, we will simply pull the defining function back to all possible holomorphic curves and compute the orders of vanishing.

Before turning to these algebraic-geometric considerations, we continue with additional differential geometric information on the Levi form. The main consideration remains the interaction between the real and the complex.

3.1.2 Almost complex structure

For some purposes it is inconvenient to complexify the tangent bundle of a smooth real submanifold M. In this section we describe some of the ideas from Section 1.1, while attempting to focus on the real tangent bundle TM. As before, we begin with geometric considerations on $\mathbb{C}^n$. The real tangent spaces satisfy $T_p\mathbb{C}^n \simeq \mathbb{C}^n \simeq \mathbf{R}^{2n}$. If we choose coordinates $z^j = x^j + iy^j$ on $\mathbb{C}^n$, then we have the coordinate vector fields $\partial/\partial x^j$ and $\partial/\partial y^j$. Fix a point p and consider the endomorphism $J : T_p\mathbb{C}^n \to T_p\mathbb{C}^n$, defined by

$$J\left(\frac{\partial}{\partial x^j}\right) = \frac{\partial}{\partial y^j}$$

$$J\left(\frac{\partial}{\partial y^j}\right) = -\frac{\partial}{\partial x^j} \tag{21}$$

where the coordinate vector fields are evaluated at p. Then $J^2 = -\mathrm{Id}$. After defining J at each point, we obtain what is sometimes called the almost complex structure endomorphism. Alternatively we can consider an endomorphism satisfying $J^2 = -\mathrm{Id}$ to be the basic object, then proceed to define the Cauchy–Riemann structure from it.

For motivation we work in local coordinates. We express the relationship between the real and complex by using J. First we write down an isomorphism between the complex vector spaces $T_p\mathbb{C}^n$ and $T_p^{1,0}\mathbb{C}^n \subset T_p\mathbb{C}^n \otimes \mathbb{C}$. Consider a real vector field

$$X = \sum_{j=1}^{n} a_j \frac{\partial}{\partial x^j} + \sum_{j=1}^{n} b_j \frac{\partial}{\partial y^j}. \tag{22}$$

To it we assign the complex vector field

$$L = \sum_{j=1}^{n} 2\left(a_j + ib_j\right)\frac{\partial}{\partial z^j}. \tag{23}$$

A simple verification reveals that

$$L = X - iJX. \tag{24}$$

The correspondence $X \to L$ is an isomorphism, with inverse mapping $L \to \text{Re}(L) = X$.

This isomorphism enables us to pass back and forth between real vector fields and type (1,0) vector fields. It also applies to vector fields on real hypersurfaces. Consider now a real hypersurface $M \subset \mathbb{C}^n$. If $X \in TM$, then $L \in T^{1,0}M$. Conversely, if $L \in T^{1,0}M$, then $\overline{L}$ is also tangent, so $X = \frac{1}{2}(L + \overline{L})$ is tangent. Therefore $X \in TM$. This observation enables us to define the Levi form on the tangent bundle, without complexification. Suppose that

$$L = X + iY = X - iJX. \tag{25}$$

Letting η be the differential form from Section 1, we compute the Levi form using Definition 2:

$$
\begin{aligned}
\lambda\left(L, \overline{L}\right) &= \left\langle \eta, \left[L, \overline{L}\right] \right\rangle \\
&= \left\langle \eta, -2i\left[X, Y\right] \right\rangle \\
&= \left\langle \eta, -2i\left[X, -iJX\right] \right\rangle \\
&= -2\left\langle \eta, \left[X, JX\right] \right\rangle.
\end{aligned}
\tag{26}
$$

Without using complexification we can also describe the "good" and "bad" directions. Given a real hypersurface, therefore, we consider the maximal complex subspaces of the tangent spaces; these tangent vectors are the "good directions." We let $hM \subset TM$ denote the bundle whose local sections are good directions. This bundle is closed under J. The bad direction can be chosen to be any remaining linearly independent vector. One choice is iT. Note that $J(iT)$ is no longer tangent to the hypersurface. We include the imaginary unit because (11) had given the bad direction as purely imaginary.

Recall that Definition 1 gave the meaning of $HM = T^{1,0}M \oplus T^{0,1}M$. We note here the simple fact that HM cannot be integrable unless hM is.

LEMMA 2
Let M be a CR manifold of hypersurface type. The bundle $HM \subset \mathbb{C}TM$ cannot be integrable unless the bundle $hM \subset TM$ is integrable.

PROOF Let X, Y be local sections of hM for which $[X, Y]$ is not such a local section. We wish to construct a (1,0) vector field L such that $\left[L, \overline{L}\right]$ is not a local section of HM, that is, $\lambda\left(L, \overline{L}\right) \neq 0$. The isomorphism (24) assigns (1,0) vectors A, B to X, Y. We compute

$$
\begin{aligned}
[X, Y] &= \frac{1}{4}\left[A + \overline{A}, B + \overline{B}\right] \\
&\equiv \frac{1}{4}\left(\lambda\left(A, \overline{B}\right) - \lambda\left(B, \overline{A}\right)\right) \bmod HM.
\end{aligned}
\tag{27}
$$

If either $\lambda\left(A, \overline{A}\right) \neq 0$ or $\lambda\left(B, \overline{B}\right) \neq 0$, then we have found the desired vector field. If both vanish, then we consider $L = A + iB$. Assuming both Levi forms vanish, we have

$$\lambda\left(A + iB, \overline{A + iB}\right) = i\lambda\left(B, \overline{A}\right) - i\lambda\left(A, \overline{B}\right). \tag{28}$$

Since we are assuming that (27) doesn't vanish, neither does (28). Therefore HM isn't integrable either. ∎

Suppose that we have an almost complex structure endomorphism on the tangent spaces of a real manifold. We can use (24) to define vector fields of type (1,0) and then define vector fields of type (0,1) by complex conjugation. The local sections do not necessarily define bundles, because the rank of the spaces need not be constant. Furthermore, even if the (1,0) vectors define a bundle, the bundle need not be integrable. These remarks indicate why we define a CR manifold as in Definition 1. It is perhaps appropriate to mention that CR manifolds with CR codimension equal to zero are also called integrable almost complex manifolds. The most common definition of such an object uses the structure endomorphism J. An integrable almost complex manifold always has a complex analytic structure; this was proved by Newlander and Nirenberg in 1957. See [We] for more information about almost complex structures and [FK] for Kohn's proof of the Newlander–Nirenberg theorem.

3.1.3   Parameterized holomorphic curves

Already in Chapter 1 we saw the usefulness of knowing that a sphere contains no parameterized holomorphic curves. Let us relate such curves to the Levi form. Suppose that z is a parameterized holomorphic curve in $\mathbb{C}^n$. We may suppose that

$$z : \{|t| < 1\} \to \mathbb{C}^n$$
$$z(0) = 0 \in M \tag{29}$$

and the real hypersurface M has local defining function r. Let us discover what happens when the curve lies in M. By differentiating the relation

$$0 = r(z(t)) = z^*r, \tag{30}$$

we obtain the relations

$$\sum_{i=1}^{n} r_i(0)\, z_i'(0) = 0 \tag{31}$$

and

$$\sum_{i,j=1}^{n} r_{i\bar{j}}(0)\, z_i'(0)\, \overline{z_j'(0)} = 0. \tag{32}$$

Equation (31) says that the derivative $z'(0)$ is a $(1,0)$ vector at 0, tangent to M. Equation (32) says that this vector is also in the kernel of the Levi form. Under the assumption that M is strongly pseudoconvex at 0, both these equations cannot hold unless $z'(0)$ vanishes. By higher differentiation, or by passing to a nonsingular point, one obtains the result that there can be no complex analytic varieties in a strongly pseudoconvex hypersurface.

Later we will generalize this argument considerably, thereby enabling us to measure how closely complex analytic varieties contact a real hypersurface. From the strongly pseudoconvex case we observe that definiteness of the Levi form is an obstruction to finding complex varieties in M. Here is a simple generalization of this argument. At a fixed point p, we may suppose that the Levi form has P positive eigenvalues, N negative eigenvalues, and $Z = n-1-P-N$ vanishing eigenvalues. We consider the quantities $\max(P, N)$ and the absolute signature of the Levi form $|s| = |P - N|$. Then the existence of complex analytic varieties in a hypersurface gives some information on the absolute signature of the Levi form and on the maximum number of eigenvalues of one sign. In Chapter 4 we prove a deeper theorem, involving higher order terms of the Taylor series of a defining function.

THEOREM 1

Suppose that M is a real hypersurface that contains a q-dimensional complex variety V. Then the following inequality must hold at all points of V:

$$|P - N| \leq \max(P, N) \leq n - 1 - q. \tag{33}$$

PROOF First we suppose that p is a smooth point of V. Then we can find a local biholomorphic mapping from a neighborhood of the origin in $\mathbb{C}^q$ to a neighborhood of p in V. Using the above notation, we denote this mapping by $t \to z(t)$. Since it is biholomorphic, the derivative $\partial z / \partial t$ has rank q. Letting r be a local defining function for M near p, we see that $z^* r = r(z(t))$ vanishes identically in t. Again we compute the first and second derivatives and evaluate at the origin. Since the mapping is holomorphic, we obtain, for $j = 1, \ldots, q$, that

$$\frac{\partial}{\partial t^j}(r(z(t))) = \sum_{k=1}^{n} \frac{\partial r}{\partial z^k} \frac{\partial z^k}{\partial t^j} = 0 \tag{34}$$

and

$$\frac{\partial^2}{\partial t^j \partial \overline{t^j}}(r(z(t))) = \sum_{k=1}^{n} \sum_{l=1}^{n} \frac{\partial^2 r}{\partial z^k \partial \overline{z^l}} \frac{\partial z^k}{\partial t^j} \frac{\partial \overline{z^l}}{\partial \overline{t^j}} = 0. \tag{35}$$

Let $(\partial z / \partial t)_0$ denote the derivative mapping evaluated at the origin. Then (34) reveals that

$$\left(\frac{\partial z}{\partial t}\right)_0 \left(\frac{\partial}{\partial t^j}\right) \in T_p^{1,0}(M). \tag{36}$$

Since the derivative has full rank, the q vectors in (36) are linearly independent $(1,0)$ tangent vectors at p. Therefore they span $T_p^{1,0}V$. Equation (35) says that the Levi form vanishes on all these vectors. We conclude the following. For each complex manifold $V = V^q \subset M$, and at each point p, it follows that

$$T_p^{1,0}V \subset T_p^{1,0}M$$

$$\lambda\left(L, \overline{L}\right) = 0 \quad \forall L \in T_p^{1,0}V. \tag{37}$$

Next suppose that V is a variety. Each point of V is a limit of smooth points, by the statement of the local parameterization theorem. Since the Levi form annihilates q independent vectors at each smooth point, it annihilates at least q independent vectors at each point: simply take limits. If $V = V^q$ is therefore a complex variety lying in a real hypersurface M, then at each point $p \in V$ there are q linearly independent $(1,0)$ vectors in the kernel of the Levi form. In order to prove the theorem, we must count the number of independent vectors annihilated by the Levi form.

We must distinguish the interpretation of the Levi form as a matrix from its interpretation as an Hermitian form. If the Levi form is assumed to be semidefinite, then the kernel of the matrix of the Levi form equals the kernel of the corresponding Hermitian form. Hence in that case the rank of the Levi form at p is at most $n - 1 - q$ and the result follows. In general we may choose coordinates so that the Levi matrix has the form

$$\begin{pmatrix} I_P & 0 & 0 \\ 0 & -I_N & 0 \\ 0 & 0 & 0_Z \end{pmatrix}. \tag{38}$$

We have $P + N + Z = n - 1$. After multiplying through by (-1) we may assume that $P \geq N$. The condition that $\lambda(v, \overline{v}) = 0$ becomes the condition that $\|I_P v\|^2 = \|I_N v\|^2$, which is not a linear condition. We can count the number of parameters. We want $\lambda(v, \overline{v}) = 0$. We may choose Z arbitrary parameters corresponding to the submatrix 0_Z. We may also pick N arbitrary parameters corresponding to I_N. To ensure that $\|I_P v\|^2 = \|I_N v\|^2$ holds, there must be a linear map U for which

$$I_P v = U I_N v \Leftrightarrow (I_P - U I_N) v = 0. \tag{39}$$

This determines the rest of the parameters and shows that the maximum number of independent vectors annihilated by the Hermitian form equals

$$q = Z + \min(P, N) = n - 1 - \max(P, N). \tag{40}$$

Statement (40) implies the conclusion of the theorem. ∎

It is perhaps worth remarking that the collection of vectors satisfying $\lambda(v, \overline{v}) = 0$ is not in general a linear subspace.

Example 1

Detecting whether there is a complex variety in a real hypersurface involves more than a glance at the Levi form. Here are two examples where the rank is full everywhere, and there is both one positive and one negative eigenvalue. The hypersurfaces are defined by the polynomials

$$r(z, \bar{z}) = 2 \operatorname{Re}(z_3) + |z_1|^2 - |z_2|^2$$
$$s(z, \bar{z}) = 2 \operatorname{Re}(z_3) + |z_1|^2 + |z_1|^4 - |z_2|^2. \tag{41}$$

The Levi forms are the matrices

$$\begin{pmatrix} 1 & 0 \\ 0 & -1 \end{pmatrix}$$

$$\begin{pmatrix} 1 + 4|z_1|^2 & 0 \\ 0 & -1 \end{pmatrix}. \tag{42}$$

The first hypersurface contains the complex line defined by

$$z_1 = z_2$$
$$z_3 = 0 \tag{43}$$

while there is no complex variety lying in the second hypersurface. ⬚

COROLLARY 1

If the Levi form on a real hypersurface has k eigenvalues of the same sign at each point, then there cannot be a complex variety of dimension $n - k$ lying in the hypersurface.

PROOF If a complex variety lying in the hypersurface has dimension q, then it follows from the theorem that

$$q \leq n - 1 - \max(P, N) \leq n - 1 - k. \tag{44}$$

This is the desired conclusion. ∎

In the next section we determine precisely which real-valued polynomials contain complex varieties in their zero sets. The method is based on algebraic geometric ideas rather than on the Levi form. We will eventually extend these ideas to the general case. It is important however to mention the following fact. If the Levi form on a smooth hypersurface has constant rank k, then the hypersurface is foliated by complex analytic varieties of dimension $n - 1 - k$. See the exercises in [Kr] for one way to reduce this statement to the Frobenius theorem. In Chapter 6 we say considerably more about how the kernel of the Levi form controls the existence of complex analytic varieties in a pseudoconvex real analytic hypersurface.

3.2   Algebraic real hypersurfaces and complex varieties

3.2.1   The associated family of holomorphic ideals

In this section we consider real hypersurfaces and more generally real subvarieties that are defined by polynomial equations. The restriction to polynomials is temporary; the ideas here will be refined in Chapter 4. For now we suppose that r is a real-valued polynomial on $\mathbb{C}^n$. We give a complete answer to whether or not a q-dimensional complex analytic variety can lie in the real subvariety defined by $\{z : r(z, \bar{z}) = 0\}$. Later we will extend this understanding to the real analytic case; we will also determine when no complex variety can have arbitrarily high contact with a smooth real subvariety. In order to handle the general (smooth) case, one approximates by Taylor polynomials, so it is advantageous to understand the polynomial case first. Furthermore the ideas in this case are transparent and evince an elegance that no generalization along the lines of Theorem 1 could ever hope to achieve.

Suppose that

$$r(z, \bar{z}) = \sum_{a,b} c_{ab}(p)(z - p)^a \left(\overline{z - p}\right)^b \tag{45}$$

is a real-valued polynomial of degree m on $\mathbb{C}^n$. We will treat the coefficients c_{ab} as an Hermitian form on a finite dimensional vector space of holomorphic polynomials, thereby reducing the problem to deciding whether certain holomorphic polynomial equations define nontrivial complex analytic varieties. Let us suppose that p lies in $\{r = 0\}$, but not assume that $dr(p)$ is nonzero. This enables us to treat the case of higher codimension real subvarieties. Since r is real-valued, we have

$$c_{ab}(p) = \overline{c_{ba}(p)}. \tag{46}$$

It is therefore possible to identify c_{ab} with an Hermitian form, then diagonalize it. The easiest way to do so is explicitly, using the polarization identity. It is also convenient to first isolate the pure terms. This leads to the definitions

$$h_p(z) = 4 \sum_{|a|=1}^{m} c_{ao}(p)(z - p)^a \tag{47}$$

$$f_p^b(z) = \sum_{|a|=1}^{m} c_{ab}(p)(z - p)^a + (z - p)^b \tag{48}$$

$$g_p^b(z) = \sum_{|a|=1}^{m} c_{ab}(p)(z - p)^a - (z - p)^b. \tag{49}$$

With these definitions, we have the useful formula

$$4r(z, \bar{z}) = 2\mathrm{Re}\left(h_p(z)\right) + \|f_p(z)\|^2 - \|g_p(z)\|^2. \tag{50}$$

There are several valuable observations about this formula, called the holomorphic decomposition of r. First of all, the real-valued polynomial r depends on both z and $\bar{z}$. By allowing squared norms and real parts, we have expressed it solely in terms of holomorphic polynomials. Second, we have isolated the "pure" terms from the "mixed" terms. We will see later that this distinction is preserved if one chooses a different defining function. Third, there is geometric intuition behind this decomposition. The pure terms define an algebraic analogue of the "bad direction." The other terms give a decomposition into pseudoconvex and pseudoconcave parts. Fourth, the condition that there be no negative squares in the decomposition is simply that the matrix $c_{ab}(p)$ be positive semidefinite. In Section 3.3 we apply the same ideas to a real analytic defining function, and give a simple way to conceive of the decomposition. In Chapter 5 we will also pass between a real-valued polynomial and its corresponding Hermitian form.

REMARK 1 For some purposes it is preferable not to isolate the pure terms. By writing

$$2\,\mathrm{Re}(h) = \left|h + \frac{1}{2}\right|^2 - \left|h - \frac{1}{2}\right|^2 \tag{51}$$

we find that every defining function r has a decomposition

$$4r = \|F\|^2 - \|G\|^2. \tag{52}$$

The disadvantage of (52) is that two of the functions involved do not then vanish at the point in question. One advantage is that it is very easy to multiply together two expressions of the form (52). Another advantage will be seen in the discussion of Example 2. Either form shows that there is a holomorphic mapping from any algebraic real hypersurface to a quadratic real hypersurface in more variables. ▮

Suppose now that there is a positive dimensional complex analytic variety Z that lies in the zero set of r. We can then choose an irreducible branch V of a one-dimensional subvariety of Z. Since V is one-dimensional, its normalization is an open subset $\Omega \subset \mathbb{C}$ and a holomorphic mapping $z : \Omega \to V$. See [Gu1,Wh] for the theorem that a variety has a normalization; the one-dimensional case is much simpler. The result needed here follows also from the local parameterization theorem. The existence of the normalization has the following consequence. If there exists a positive dimensional complex analytic variety Z that lies in $\{r = 0\}$ and passes through p, then there is a nonconstant mapping

$$z : (\mathbb{C}, 0) \to (\mathbb{C}^n, p) \tag{53}$$

that satisfies

$$z^* r = r(z(t)) = 0. \tag{54}$$

Substitution into the holomorphic decomposition yields

$$0 = 4z^* r = 2 \operatorname{Re} \left(z^* h \right) + ||z^* f||^2 - ||z^* g||^2. \tag{55}$$

One can differentiate this equation and obtain an infinite system of equations for the coefficients of the parameterized curve; it is expeditious to first work on the pure terms. We claim that $z^* h$ vanishes. To see this, differentiate repeatedly with respect to t while treating $\bar{t}$ as an independent variable. Note that all the terms in $||z^* f||^2 - ||z^* g||^2$ are divisible by $\bar{t}$ and hence vanish when we set $t = \bar{t} = 0$. This yields

$$0 = \left(\frac{\partial}{\partial t} \right)^s \left(z^* r \right) (0) = \left(\frac{\partial}{\partial t} \right)^s \left(z^* h \right) (0) \quad \forall s. \tag{56}$$

Hence the holomorphic function $z^* h$ has trivial Taylor series at 0 and therefore vanishes identically. It then follows that $||z^* f||^2 = ||z^* g||^2$, and we have an equality of squared norms of holomorphic vector-valued functions. This suggests consideration of the group $U(N)$ of unitary matrices on $\mathbb{C}^N$. The following proposition suffices for understanding this equality of squared norms, although later we require a more general version.

PROPOSITION 3
Suppose that B is an open ball about 0 in $\mathbb{C}^q$, and F, G are holomorphic mappings from B to $\mathbb{C}^N$ for which

$$||F||^2 = ||G||^2. \tag{57}$$

Then there exists $U \in U(N)$ for which $F = UG$.

PROOF   We may suppose that

$$F(z) = \sum F_a z^a \tag{58}$$

$$G(z) = \sum G_a z^a \tag{59}$$

have convergent power series in B, with coefficients F_a and G_a that are elements of $\mathbb{C}^N$. By expanding and equating the norms squared and writing $(\ ,\ )$ for the inner product on $\mathbb{C}^N$, we obtain the relations

$$(F_a, F_b) = (G_a, G_b) \tag{60}$$

for all choices of multi-indices a, b. Choose now a maximal linearly independent set among the G_a and define U on that set by

$$U G_a = F_a. \tag{61}$$

Extend by linearity. It is necessary to verify that U is well defined on the span of these G_a. Suppose that we have a dependence relationship of the form

$$G_k = \sum_j c_k^j G_j. \tag{62}$$

Then F_k has two definitions. To verify that U is well defined, we must check that the two formulas

$$F_k = UG_k \stackrel{?}{=} U \sum_j c_k^j G_j = \sum_j c_k^j UG_j = \sum_j c_k^j F_j \tag{63}$$

agree. To do so, compute

$$\left\| F_k - \sum c_k^j F_j \right\|^2 = \left\| G_k - \sum c_k^j G_j \right\|^2 = 0 \tag{64}$$

where the middle equality follows after expansion of each squared norm by the equality of each $\left(F_j, F_k \right) = \left(G_j, G_k \right)$. Hence U is a well-defined linear transformation that preserves the inner product of every pair of vectors in its domain. Thus U is an isometry from the span $\{G_a\}$ to the span $\{F_a\}$. By defining U to be the identity on the orthogonal complement of this span, we obtain a unitary transformation of $\mathbb{C}^N$. ∎

Returning to the previous situation (55), with

$$F = z^* f$$
$$G = z^* g \tag{65}$$

we obtain an element U of $\mathbf{U}(N)$, depending on the parameterized holomorphic curve, for which $z^* f = U z^* g$. Since U is independent of the parameter t, we conclude the following:

$$0 = z^* r = 2 \operatorname{Re} \left(z^* h \right) + \| z^* f \|^2 - \| z^* g \|^2 \Rightarrow$$
$$z^* h = 0 \text{ and } \exists U : z^* (f - Ug) = 0. \tag{66}$$

This is the motivation for the important Definition 4 and Theorem 2.

The case for higher dimensional varieties is similar. By the local parameterization theorem, if $V = V^q$ is an irreducible q-dimensional complex analytic subvariety of $\mathbb{C}^n$, then there is a neighborhood Ω of the origin in $\mathbb{C}^q$, a thin set E, and a holomorphic mapping $z : \Omega - E \to V \subset \mathbb{C}^n$ whose derivative is generically of rank q. As we noted in Chapter 1, we cannot generally find surjective parameterizations of the variety. By pulling back to a mapping of generic rank q one sees that

$$\| f(z) \|^2 = \| g(z) \|^2 \; \forall z \in V \Rightarrow \exists U : f(z) = Ug(z) \; \forall z \in V. \tag{67}$$

The holomorphic decomposition is a simple idea. One can conceive of it either as resulting from the diagonalization of an Hermitian form or from the polarization identity

$$|z + w|^2 + |z - w|^2 = 4\mathrm{Re}\,(z\overline{w})\,. \tag{68}$$

Despite its simplicity, this idea has profound consequences for the geometry of real and complex objects. In some sense the rest of this book is devoted to a description of these consequences. The ideals arising from a holomorphic decomposition will arise often, so we name them.

DEFINITION 4   *Suppose that M is a real algebraic subvariety of $\mathbb{C}^n$ and that*

$$4r_p = 2\,\mathrm{Re}\,(h^p) + ||f^p||^2 - ||g^p||^2 \tag{69}$$

is a holomorphic decomposition of any polynomial defining function for M at some point p. Let N be the maximum number of components of f^p and g^p. For $U \in \mathbf{U}(N)$, we consider the ideals

$$I\,(U,p) = (h, f - Ug) \tag{70}$$

in O_p generated by h^p and the components of $f^p - Ug^p$. The family of ideals $I\,(U,p)$ is called the associated family of holomorphic ideals on M.

REMARK 2   Given a holomorphic decomposition, we may need to include some functions that are identically zero to ensure that there are the same number of f's as g's. Suppose that this number is N. We could have also elected to write the decomposition without isolating the pure terms, say as $||F||^2 - ||G||^2$, by writing

$$2\,\mathrm{Re}\,(h) = \left|h + \frac{1}{2}\right|^2 - \left|h - \frac{1}{2}\right|^2\,.$$

The number of component functions would then be $N+1$, so we would need to consider matrices from $\mathbf{U}(N+1)$. Each element of $\mathbf{U}(N)$ gives one on $\mathbf{U}(N+1)$ in an obvious manner. The extra unitary matrices create functions in the ideal that do not vanish; the ideal $I\,(U,p)$ is therefore the whole ring for these unitary matrices. Since these define empty varieties, there is no difference in the theory if we write the holomorphic decomposition without isolating the pure terms. ∎

In the section on real analytic subvarieties, we will prove that the associated family is independent of the defining function or choice of decomposition. The individual ideals depend on the defining function, but the collection of them does not.

It is obvious that the varieties $\mathbf{V}\,(I\,(U,p))$ lie in M; the main point of this chapter has been establishing the converse. We state this in Theorem 2 below.

Let us note here an interesting change of point of view. Even when all these varieties are trivial, the defining equation still reflects the ideals that define them. One can think of each ideal as defining a zero-dimensional scheme that lies in M. This point of view has important consequences for proper mappings. As in algebraic geometry, it is not just the point set defined by these functions that matters, but also the ideal defined by them. See [Sfv] for the definition of a scheme.

It is appropriate to discuss the extent to which these choices are unique. For a fixed polynomial, the holomorphic decomposition is not unique. The function h is uniquely determined; it follows that $\|f\|^2 - \|g\|^2$ is also. This does not determine the components. It is a consequence of Proposition 3 that

$$\|f\|^2 - \|g\|^2 = \|F\|^2 - \|G\|^2 \Rightarrow \|f\|^2 + \|G\|^2 = \|g\|^2 + \|F\|^2$$

$$\Rightarrow \begin{pmatrix} f \\ G \end{pmatrix} = \begin{pmatrix} u_{11} & u_{12} \\ u_{21} & u_{22} \end{pmatrix} \begin{pmatrix} g \\ F \end{pmatrix} \tag{71}$$

where the matrix (of matrices)

$$U = \begin{pmatrix} u_{11} & u_{12} \\ u_{21} & u_{22} \end{pmatrix} \tag{72}$$

is unitary. It turns out that the ambiguity is accounted for completely by considering the whole collection of ideals.

Theorem 2 below summarizes the discussion; it completely answers the question of whether there is a complex analytic variety in the zero set of a real-valued polynomial. Our aim in Chapter 4 will be to make this theorem quantitative and thereby afford a method for applying it to the Taylor polynomials of a smooth function.

THEOREM 2
Let M be the real algebraic subvariety defined by the vanishing of a real-valued polynomial r on $\mathbb{C}^n$. For each point p and each unitary transformation U, let $I(U, p) = (h^p, f^p - Ug^p)$ denote the associated family of holomorphic ideals. Then

$$\mathbf{V}(I(U, p)) \tag{73}$$

contains p and lies in M for every U. Conversely, every irreducible complex analytic subvariety of $\mathbb{C}^n$ that lies in M and passes through p must be a subvariety of some

$$\mathbf{V}(I(U, p)). \tag{74}$$

We obtain immediately the following corollaries.

COROLLARY 2

There is no positive dimensional complex analytic variety V lying in a compact real algebraic subvariety M of $\mathbb{C}^n$.

PROOF   Without loss of generality we may suppose that the variety is irreducible. According to the theorem, if there is such a complex variety lying in M, then it is a subvariety of some

$$\mathbf{V}\left(I\left(U,p\right)\right), \tag{75}$$

and this algebraic variety lies in M also. Since this variety is given by polynomial equations, it is globally defined. It also stays within M. But it is a consequence of the maximum principle that there is no compact positive dimensional complex analytic subvariety of $\mathbb{C}^n$.   ∎

COROLLARY 3

Suppose that

$$r\left(z,\bar{z}\right) = 2\,\mathrm{Re}\left(h\left(z-p\right)\right) + \sum_{|a|,|b|>0} c_{ab}\left(p\right)\left(z-p\right)^{a}\left(\overline{z-p}\right)^{b} \tag{76}$$

is a real-valued polynomial. Then there must be positive dimensional complex analytic varieties through p and lying in $M = \{r = 0\}$ whenever the Hermitian matrix of coefficients satisfies

$$\mathrm{rank}\left(c_{ab}\left(p\right)\right) < n - 1$$

$$|a|, |b| > 0. \tag{77}$$

PROOF   The condition on the rank ensures that the number of f's and the number of g's is at most $n-2$. The function h is in the associated ideal, so the associated ideal has at most $n-1$ generators. Thus, for every U, $\dim\left(\mathbf{V}\left(I\left(U,p\right)\right)\right) > 0$.   ∎

COROLLARY 4

There is a q-dimensional complex analytic variety through a point p and lying in M if and only if there is a unitary matrix U for which

$$\dim\left(\mathbf{V}\left(I\left(U,p\right)\right)\right) \geq q. \tag{78}$$

PROOF   If there were such a variety, then its irreducible branches would also lie in M, so the result follows from the theorem.   ∎

Example 2

For the hypersurface defined by

$$s(z, \bar{z}) = 2 \operatorname{Re}(z_3) + |z_1|^2 + |z_1|^4 - |z_2|^2 \tag{79}$$

the associated family of ideals at the origin is easily computed to be

$$\left(z_3, z_1 - az_2, z_1^2 - cz_2\right) \tag{80}$$

where the constants satisfy $|a|^2 + |c|^2 = 1$. The variety defined by these ideals consists of the origin alone, so there is no variety lying in the hypersurface and passing through the origin.

More generally, we can check that there is no variety passing through any other point as well. To do so, we write

$$\begin{aligned}
s(z, \bar{z}) &= \left| z_3 + \frac{1}{2} \right|^2 + |z_1|^2 + |z_1^2|^2 - \left| z_3 + \frac{1}{2} \right|^2 - |z_2|^2 \\
&= \|F\|^2 - \|G\|^2.
\end{aligned} \tag{81}$$

Here we have not separated the pure terms. The method of Theorem 2 yields a family of ideals $(F - UG)$ in the polynomial ring. These global associated ideals are

$$\left(z_3 + \frac{1}{2} - a\left(z_3 - \frac{1}{2}\right) - bz_2, \, z_1 - d\left(z_3 - \frac{1}{2}\right) - ez_2, \, z_1^2 - g\left(z_3 - \frac{1}{2}\right) - hz_2 \right) \tag{82}$$

where

$$\begin{pmatrix} a & b & c \\ d & e & f \\ g & h & k \end{pmatrix} \in \mathbf{U}(3).$$

The resulting three generators define a positive dimensional variety for no choice of this unitary matrix, so the result follows.   □

Example 3

Let M be the hypersurface defined by

$$2 \operatorname{Re}(z_4) + |z_1|^2 + |z_2|^2 - |z_3|^2. \tag{83}$$

Each of the complex lines

$$\begin{aligned}
V_1 &= \{(t, 0, t, 0) : t \in \mathbb{C}\} \\
V_2 &= \{(0, t, t, 0) : t \in \mathbb{C}\}
\end{aligned} \tag{84}$$

lies in M and hence so does their union. Their union is not, however, a subvariety of the variety

$$\mathbf{V}\left(I\left(U,0\right)\right) = \mathbf{V}\left(z_4, z_1 - az_3, z_2 - cz_3\right)$$

$$|a|^2 + |c|^2 = 1 \tag{85}$$

for any choice of a unitary matrix

$$\begin{pmatrix} a & b \\ c & d \end{pmatrix}.$$

This shows that the assumption of irreducibility in the statement of Theorem 2 is required. Each irreducible branch in this example determines a different unitary matrix. One way to view this is as follows. For each point

$$\xi = (a, c) \in S^3 \subset \mathbb{C}^2 \tag{86}$$

there is a one-dimensional complex line V_ξ lying in the hypersurface M. Since the sphere has no complex analytic structure within it, it is impossible to piece these lines into a higher dimensional variety. $\Box$

Yet another corollary is a theorem of Bezout type. Recall [Sfv] that the classical Bezout theorem tells us about the number of intersections of homogeneous polynomial equations in projective space. If the loci have a common component, then there are infinitely many solutions. If they have no common component, then the number of solutions (with multiplicities properly counted) equals the product of the degrees of the defining polynomials. In our situation there is a similar alternative. Suppose we wish to know whether there is a complex analytic variety in a real subvariety of complex n-space. It is a consequence of Theorem 6 that any complex analytic variety with sufficiently high order of contact with a real algebraic subvariety must actually lie within it. The bound depends only on the dimension and the degree of the defining polynomial. Careful statements of such results in the smooth case require making our results quantitative. This will be accomplished in Chapter 4.

3.3   Real analytic subvarieties

3.3.1   Holomorphic decomposition

Before turning to numerical measurements of the contact of complex varieties with real algebraic subvarieties, we discuss the real analytic case. The real analytic case is essentially the same as the polynomial case, although convergence proofs are required. Furthermore, it is easy to prove that the associated family of holomorphic ideals is invariantly associated with a real analytic real subvariety, thereby rendering it unnecessary to verify this separately in the algebraic case.

The analogues of all the theorems in the algebraic case remain true except for the Bezout theorem. As there is no notion of degree for a power series, this is not surprising. A uniform boundedness principle is true, however, in the real analytic case. This can be phrased as the following alternative. Either there is a complex analytic variety in a real analytic subvariety through a given point, or the point is a point of finite type. See Definition 6 for the notion of finite type.

In order to prove these statements, we need to treat the coefficients of a defining function as an Hermitian form on an infinite dimensional space. It is convenient to begin with these considerations. Suppose that

$$f^j : \Omega \to \mathbb{C}, \quad j = 1, 2, \ldots \tag{87}$$

is a countable collection of holomorphic functions on a domain Ω. If the infinite sum

$$s(z, \bar{z}) = \sum_{j=1}^{\infty} \left| f^j(z) \right|^2 \tag{88}$$

converges uniformly on compact subsets of Ω, then it will represent a real analytic function there. It is convenient to consider holomorphic functions with values in a Hilbert space. We write

$$f : \Omega \to l^2 \tag{89}$$

to indicate that

$$\|f\|^2 = \sum_{j=1}^{\infty} \left| f^j \right|^2 \tag{90}$$

is square summable. In case the domain is a polydisc centered at the origin, a holomorphic function is always represented by a power series about the origin that converges in the whole polydisc. This assertion follows immediately from the one-dimensional case. It is therefore possible to identify

$$f : \Omega \to l^2 \tag{91}$$

with the countable collection of sequences defined by the Taylor coefficients. Thus

$$f^j = \sum f_a^j z^a \tag{92}$$

and we also write

$$f = \sum f_a z^a \tag{93}$$

where now the coefficients are themselves sequences. They are also square summable. This follows from the Cauchy estimates:

$$\left| f_a^j \right| \leq r^{-a} \left\| f^j \right\|_{P_r}, \tag{94}$$

where r is any multi-radius inside the polydisc of definition. Summing both sides of (94) on j, we see that f_a also defines an element of l^2. We therefore identify the original countable collection of square summable holomorphic functions with a countable collection of elements in l^2 defined by the Taylor coefficients.

These remarks aid us in the holomorphic decomposition of a real-valued, real analytic function. Suppose now that

$$r(z, \bar{z}) = \sum_{a,b} c_{ab} (z - p)^a \overline{(z - p)^b} \tag{95}$$

is a real analytic function, where the power series converges on the polydisc $P = P_\delta(p)$ given by $P = \{z : |z^j - p^j| < \delta_j\}$. We wish to decompose r as we did before, but into holomorphic functions rather than simply holomorphic polynomials. The proof is similar, but we must pay attention to convergence. Let t be a positive real number less than one. Define holomorphic functions h, f, g as follows:

$$h_p(z) = 4 \sum_a c_{ao}(p)(z - p)^a \tag{96}$$

$$f_p^b(z) = \sum_a c_{ab}(p)(t\delta)^b (z - p)^a + (z - p)^b (t\delta)^{-b} \tag{97}$$

$$g_p^b(z) = \sum_a c_{ab}(p)(t\delta)^b (z - p)^a - (z - p)^b (t\delta)^{-b}. \tag{98}$$

The basic formula

$$4r(z, \bar{z}) = 2 \operatorname{Re}\left(h_p(z)\right) + ||f_p||^2 - ||g_p||^2 \tag{99}$$

still holds, where h is holomorphic and the norms-squared define real analytic functions. We verify this next. The formula for h is obvious, as it follows from setting $\overline{z - p} = 0$ in the definition of r. (Again the idea of treating a variable and its conjugate as independent variables appears.) The formula obviously works formally, so it remains only to verify the convergence of the norms-squared. Because of the elementary inequality

$$|x + y|^2 \leq 2\left(|x|^2 + |y|^2\right), \tag{100}$$

it is enough to verify that

$$\sum_b \left| \sum_a c_{ab}(p)(t\delta)^b (z - p)^a \right|^2 \tag{101}$$

and

$$\sum_b \left| (z - p)^b (t\delta)^{-b} \right|^2 \tag{102}$$

are summable. It is sufficient to verify that each sum converges without the square. Removing the squares, we must verify the convergence of

$$\sum_b \left| \sum_a c_{ab}(p)(t\delta)^b (z-p)^a \right| \tag{103}$$

and

$$\sum_b \left| (z-p)^b (t\delta)^{-b} \right|. \tag{104}$$

The second sum is a multiple geometric series that converges in the polydisc $\left| z^j - p^j \right| < t\delta_j$. The first is no more than the absolute evaluation

$$\sum_b \sum_a \left| c_{ab}(p)(t\delta)^b (z-p)^a \right| \tag{105}$$

of the original defining function within its polydisc of convergence, because of our choice of t. Since the power series converges absolutely and uniformly on compact subsets of $P_\delta(p)$, so does (105). Thus both series converge and the squared norms define real analytic functions.

In this manner we associate with a real analytic function r the holomorphic function h and the holomorphic l^2-valued functions f, g. As before, we will need to know what happens when the squared norms of these vector-valued functions are the same.

PROPOSITION 4
Let

$$F, G : \Omega \to l^2$$

$$\|F\|^2 = \|G\|^2 \tag{106}$$

be holomorphic mappings with the same squared norms. Let F_a, G_a denote the element of l^2 determined as the coefficients of the power series:

$$F = \sum F_a z^a$$
$$G = \sum G_a z^a. \tag{107}$$

Then there is a unitary operator $U : \text{span}\{G_a\} \to \text{span}\{F_a\}$ such that

$$F = UG. \tag{108}$$

PROOF As before, we define the operator $U : \text{span}\{G_a\} \to \text{span}\{F_a\}$ by decreeing that

$$UG_a = F_a. \tag{109}$$

Then the same argument as in Proposition 4 shows that U is well-defined and can be extended to the whole space to be unitary. ∎

As in the polynomial case, there is no unique holomorphic decomposition for a fixed defining function. Furthermore, it is necessary that our concepts be invariant under the choice of a new defining function. By considering the associated family of holomorphic ideals, one can account for both possible difficulties.

DEFINITION 5 *Let M be a real analytic real subvariety of* $\mathbb{C}^n$. *The associated family of holomorphic ideals at a point p in M is the family of ideals*

$$I(U,p) = (h^p, f^p - Ug^p) \tag{110}$$

in the ring O_p *defined by the holomorphic decomposition of any defining function. We assume that* U : span $\{g_a\} \to$ span $\{f_a\}$ *is unitary.*

Before we verify that this family of ideals is invariantly associated with a real analytic subvariety, we give a formal calculation that abbreviates the definitions of the functions h^p, f^p, g^p and of the ideals $I(U,p)$.

Think of the expression $(z - p)$ as an element of an infinite dimensional Hilbert space with inner product $\langle , \rangle$. This space has a complete orthonormal set $\{e_\alpha\}$, indexed by the nonzero multi-indices. Then, $(z - p)$ is an abbreviation for the vector whose inner product with e_α is the monomial $(z - p)^\alpha$. The power series for the defining equation has Taylor coefficients $c_{\alpha\beta}(p)$. After removing the pure terms, the expression

$$\sum c_{\alpha\beta}(p)(z-p)^\alpha \overline{(z-p)}^\beta \tag{111}$$

is the value of an Hermitian form applied to the vector $(z - p)$. The power series for r defines the matrix $C(p)$ of this Hermitian form by

$$r_p = 2 \operatorname{Re}(h^p(z-p)) + \langle C(p)(z-p),(z-p)\rangle. \tag{112}$$

We diagonalize the form by writing

$$\langle Cw, w\rangle = \left\|(tC + t^{-1}I)w\right\|^2 - \left\|(tC - t^{-1}I)w\right\|^2 \tag{113}$$

where the norm is the obvious one. We have written C for $C(p)$; we have written w for $(z - p)$, which abbreviates as above the collection of all $(z - p)^a$. The real parameter t is required in the convergence proof. The letter I denotes the identity mapping. The ideals $I(U,p)$ have the elegant representation as

$$I(U,p) = \left(h, tC + t^{-1}I - U\left(tC - t^{-1}I\right)\right)$$
$$= \left(h, tC(I - U) + t^{-1}(I + U)\right). \tag{114}$$

It is possible to view this as a linear fractional Cayley transformation. Given the Hermitian form C, we form the new mapping

$$U_C = \left(t^2 C + \mathrm{I}\right) \left(t^2 C - \mathrm{I}\right)^{-1}. \tag{115}$$

If there is a holomorphic curve along which U_C is unitary, then this holomorphic curve lies in our given real analytic subvariety. The procedure that defines f, g amounts to nothing more than diagonalizing the form C.

REMARK 3   Setting $U = -\mathrm{I}$ in (114) shows that the associated family of ideals always includes the ideal generated by the functions defined by the Hermitian form C. The condition that this particular ideal define a trivial variety is also significant. We give now a brief explanation.   ∎

Suppose that $f : \left(M', p'\right) \rightarrow (M, p)$ is a smooth CR diffeomorphism between (germs of) real analytic real hypersurfaces. A regularity theorem of Baouendi, Jacobowitz, and Treves [BJT] states that such a mapping must be itself real analytic in case the germ (M, p) is "essentially finite." This finiteness condition is roughly that the ideal generated by the functions defined by $C = C(p)$ (and also the pure term) define a trivial variety. More precisely, suppose that the power series for r_p defining the target hypersurface germ can be written

$$r_p(z, \bar{z}) = \sum c_{\alpha\beta}(p)(z - p)^\alpha \overline{(z - p)}^\beta$$

$$= 2 \operatorname{Re} \sum_{|\alpha| \geq 1} c_{\alpha o}(z - p)^\alpha + \sum_{|\beta| \geq 1} \sum_{|\alpha| \geq 1} c_{\alpha\beta}(p)(z - p)^\alpha \overline{(z - p)}^\beta$$

$$= 2 \operatorname{Re}\left(h(z - p)\right) + \sum_\beta h_\beta(z - p)\overline{(z - p)}^\beta$$

$$+ \operatorname{Im}\left(h(z - p)\right) \Phi\left(z - p, \overline{z - p}\right). \tag{116}$$

Here the second term in the last line is independent of h, the functions h_β are holomorphic, and Φ is defined by the equation. We first consider the ideal in A_p generated by $h, \bar{h}$. We work modulo this ideal, thereby enabling us to ignore the third term. The germ (M, p) is essentially finite if the ideal

$$\left(h, h_\beta\right) \tag{117}$$

(including all the h_β) in O_p defines the trivial variety consisting of p alone. Note that we may always take h to be a coordinate function.

Essential finiteness differs slightly from the statement that the ideal obtained by setting $U = -\mathrm{I}$ defines a trivial variety. The statements are precisely equivalent if the defining equation is independent of $\operatorname{Im}(h)$. If the defining equation does depend on this expression, then we must modulo out by $\left(h, \bar{h}\right)$ first to ensure that the resulting concept is invariant under multiplication by a unit. The associated family of holomorphic ideals is an invariant, according to Proposition 5 of the next section, but individual ideals in it are not. This implies that there is no precise correspondence between essential finiteness and the statement in Remark 3. The case where the defining equation is independent of $\text{Im}(h)$ is called "rigid" in [BR1]. In the rigid case, the two concepts are the same; in the general case, one must make sure to modulo out by $(h, \overline{h})$ first before defining the Hermitian form $C(p)$. See [D8] for more details on this point of view, see [Bg,Fo3] for more information on matters concerning holomorphic extension of CR functions, and see [BR2] for generalizations to the smooth and formal cases. The common thread is that an important geometric or analytic condition achieves its precise formulation by considering whether or not certain ideals define trivial varieties.

3.3.2 Changing the defining equation

Let us investigate the behavior of the holomorphic decomposition under multiplication of the defining function by a real analytic unit. First suppose that

$$r = 2 \, \text{Re}(h) + e \tag{118}$$

where h is holomorphic and e consists of only mixed terms. Suppose that

$$v = c + 2 \, \text{Re}(a) + B \tag{119}$$

is a unit in the ring of real-valued, real analytic functions, also decomposed into a constant term, a pure term, and the mixed terms. Multiplying out, and collecting again in this manner, show that

$$rv = 2 \, \text{Re}(h(c + a)) + h\overline{a} + \overline{h}a + ev$$
$$= 2 \, \text{Re}(H) + E. \tag{120}$$

Since c is a nonzero constant, and a vanishes at the origin, we see that the function $H = h(c + a)$ is a unit times h. The ideal (h) is therefore invariantly associated with the zero set of r at the origin. Working modulo this ideal, we see that E is a unit times e. As a consequence of these facts, we have the following

PROPOSITION 5
Let r be a real analytic defining function for a real analytic subvariety M in a neighborhood of the point 0 in M. Write r according to its holomorphic decomposition as

$$r = 2\text{Re}(h) + ||f||^2 - ||g||^2. \tag{121}$$

Suppose that $r^{\#}$ is another defining function with holomorphic decomposition

$$r^{\#} = 2 \, \text{Re}(H) + ||F||^2 - ||G||^2. \tag{122}$$

Then the family of ideals

$$I(U,0) = (h, f - Ug) \qquad (123)$$

as U varies over the unitary operators is the same as the family

$$I^{\#}(U,0) = (H, F - UG). \qquad (124)$$

Therefore, the associated family of holomorphic ideals is an invariant of M.

PROOF Let v be the unit defined by (119). There is no loss of generality in assuming that c is unity. We have already seen that the ideals

$$(h) = (H) \qquad (125)$$

are the same. To finish the proof we must show that each $I(U,0)$ coming from the holomorphic decomposition of r corresponds to one coming from vr, and conversely. Since the problem is symmetric in r and vr, it is sufficient to do just one implication. We can work modulo (h). Without loss of generality we write the unit $v = 1 + \|P\|^2 - \|Q\|^2$; note that we can incorporate the pure terms in this way as long as we allow the terms P, Q to be nonzero at the origin. Multiplying out and collecting terms yields

$$\begin{aligned}
\|f\|^2 - \|g\|^2 &= \left(1 + \|P\|^2 - \|Q\|^2\right)\left(\|F\|^2 - \|G\|^2\right) \\
&= \|F\|^2 + \|F \otimes P\|^2 + \|G \otimes Q\|^2 - \|G\|^2 \\
&\quad - \|G \otimes P\|^2 - \|F \otimes Q\|^2.
\end{aligned} \qquad (126)$$

In (126) we have used the tensor product notation in the usual manner:

$$\|F \otimes P\|^2 = \sum_{i,j} |F_i P_j|^2. \qquad (127)$$

Consider the unitary matrix

$$U^{\#} = \begin{pmatrix} U & 0 & 0 \\ 0 & U & 0 \\ 0 & 0 & U^* \end{pmatrix}. \qquad (128)$$

The notation is meant to imply that there is one copy of U corresponding to G, one copy for each component of P in $G \otimes P$, and one copy of U^* for each component of Q in $F \otimes Q$. A simple computation shows that $F - U^{\#}G = f - Ug$, where the equality is as ideals. For each U there is a corresponding $U^{\#}$. Hence we have containment, and therefore equality. We conclude that the corresponding families of ideals are the same:

$$\left\{I^{\#}(U,0) : U \text{ is unitary}\right\} = \left\{I(U,0) : U \text{ is unitary}\right\}. \qquad (129)$$

Thus the associated family of ideals of a real analytic real subvariety is invariantly associated with the subvariety. ∎

3.3.3   Existence of complex varieties in real analytic subvarieties

The primary concern in this book is the interaction between the real subvariety
M and ambient complex analytic varieties. The associated family of ideals

$$I(U,p) \tag{130}$$

is what enables us to use the algebraic ideas from Chapter 2. To see whether
there is a parameterized holomorphic curve through a point p in M, we need
only to decide whether there is an ideal $I(U,p)$ that defines a positive dimensional variety. To study the order of contact of parameterized curves, it will be
sufficient to study the order of contact of these ideals in the ring $O = O_p$.

As in the polynomial case, it is possible to determine completely when a real
analytic subvariety contains any complex analytic varieties. The next theorem
gives the solution. Its importance becomes apparent in Chapter 6. Theorems 3
and 4 of this chapter are two of the fundamental results in the study of domains
with real analytic boundaries.

THEOREM 3
*Let M be a real analytic subvariety of $\mathbb{C}^n$. Suppose that V is an irreducible
complex analytic subvariety passing through p and lying in M. Then there is
a unitary matrix U such that V is a subvariety of some $I(U,p)$. In particular,
if M is defined by a function whose Taylor expansion is valid in the polydisc
P, and t is any positive number less than one, then V can be extended to be a
subvariety of the polydisc tP. Conversely, every complex analytic subvariety of
$\mathbf{V}(I(U,p))$ lies in M and passes through p.*

PROOF   We have already completed most of the work. Since we have shown
that the associated family of holomorphic ideals is invariantly associated with the
germ of M at p, it is possible to choose any defining function for M and work
with the specific generators of the ideal arising from the holomorphic decomposition of that function. It is clear that every subvariety of $\mathbf{V}(I(U,p))$ lies in M.
Now suppose that V passes through p and lies in M. For any irreducible branch
of a one-dimensional subvariety of V, we can find a parameterized holomorphic
curve

$$z : (\mathbb{C}, 0) \to (\mathbb{C}^n, p) \tag{131}$$

such that

$$z^* r = 2 \operatorname{Re}\left(z^* h\right) + \|z^* f\|^2 - \|z^* g\|^2 = 0. \tag{132}$$

As in the polynomial case, we can differentiate this equation with respect to the
parameter t and obtain

$$\left(\frac{\partial}{\partial t}\right)^s \left(z^* h\right)(0) = 0 \quad \forall s. \tag{133}$$

Since h is holomorphic, it vanishes identically. This together with (132) implies that

$$||z^* f||^2 = ||z^* g||^2. \tag{134}$$

According to Proposition 4, we can find a unitary matrix such that $z^* f = U z^* g \Rightarrow z^* (f - Ug) = 0$. This implies the desired conclusion. Since the functions h, f, g are all defined on the polydisc tP and are completely determined by their Taylor coefficients at p, the equality $f - Ug = 0$ persists on all of tP. This finishes the proof in case V is one-dimensional.

Suppose now that V is irreducible and of dimension q. Let us recall the notation of the local parameterization theorem. There is a neighborhood of the origin in $\Omega \subset \mathbb{C}^q$ and a thin set E such that any of the inverses of the projections

$$\pi_j^{-1} : \Omega - E \to V - \pi_j^{-1}(E) \tag{135}$$

are holomorphic mappings to a subset of V consisting of smooth points. By pulling back to any of these mappings and applying the technique above, we obtain for each j a unitary matrix U^j such that

$$f = U^j g \text{ on Image} \left(\pi_j^{-1}(\Omega - E) \right). \tag{136}$$

Since this image is dense and f, g are holomorphic in a neighborhood of the origin, the equality persists on E also. All the U^j can be chosen to be the same, because $U^j g - U^k g = f - f = 0$ and because the complement of $\pi_j^{-1}(E)$ is connected. Therefore any irreducible variety gives rise to a corresponding unitary matrix. ∎

Next we turn to the uniform boundedness principle in the real analytic case. The theorem is similar to the idea in Bezout's theorem for polynomial equations. Given a real analytic function, there may be complex analytic varieties in its zero set. If there are not, however, then they have at most bounded order of contact. We formulate this principle in several ways. Notice that there are no pseudoconvexity assumptions in the theorem. Related, but weaker, theorems are well known for hypersurfaces that bound pseudoconvex domains. These results pass through the machinery of the $\bar{\partial}$-Neumann problem. See Chapter 6.

The uniform boundedness principle follows from the equivalences in Theorem 4. Among the equivalent conditions is the statement that p is a point of "finite type." The study of such points is the object of Chapter 4. We give the definition now, but postpone full discussion until that chapter. There we prove that statement 1 in Theorem 4 is equivalent to the other statements; at this point we can prove that the other five statements are all equivalent. First we give precise definitions of order of contact and multiplicities on real hypersurfaces.

DEFINITION 6 *Suppose that M is a smooth real hypersurface that contains the point p. The 1-type at p, written $\Delta^1(M, p)$, is the maximum order of contact* of ambient one-dimensional complex analytic varieties with M at p. It is defined by $\Delta^1(M,p) = \mathbf{T}(r_p)$. Here r_p denotes the ideal of germs of smooth functions at p and vanishing on M, and the invariant $\mathbf{T}$ was defined in Definition 2.8. We say that p is a point of finite type if $\Delta^1(M,p)$ is finite. (Sometimes we say finite 1-type to emphasize that we are studying the contact of one-dimensional varieties.)

From Chapter 2 we expect that there are other ways to measure the order of contact. The success of the codimension operator in describing singularities and our work in Chapter 4 motivate the notion of multiplicity of a point on a real hypersurface. We can now give the definition in the real analytic case. It will be extended to the smooth case in the next chapter.

DEFINITION 7 *Let M be a real analytic real subvariety of $\mathbb{C}^n$, with associated family of ideals $I(U,p)$. The multiplicity $B(M,p) = B^1(M,p)$ (also called the 1-multiplicity) of the point p on M is the number*

$$B(M,p) = 2\sup_U \mathbf{D}(I(U,p)) = 2\sup_U \left(\dim(I(U,p))/O_p\right). \qquad (137)$$

The usefulness of the multiplicity is evident in the proof of the following basic theorem.

THEOREM 4
Let M be a real analytic real subvariety of C^n, and let $I(U,p)$ be the associated family of ideals of holomorphic functions. The following are equivalent:

1. *The point p is a point of finite type.*
2. *For each unitary U,*

$$\mathbf{D}(I(U,p)) < \infty. \qquad (138)$$

3. *There is a constant c so that*

$$\mathbf{D}(I(U,p)) < c \quad \forall U. \qquad (139)$$

4. *For each unitary U,*

$$\mathbf{T}(I(U,p)) < \infty. \qquad (140)$$

5. *There is a constant c so that*

$$\mathbf{T}(I(U,p)) < c \quad \forall U. \qquad (141)$$

6. *There is no germ of a positive dimensional complex analytic variety passing through p and lying in M.*

PROOF The equivalence of statements 2 and 4 follows from inequality (136) of Chapter 2, as does the equivalence of statements 3 and 5. It is trivial that statement 3 implies statement 2. Thus the equivalence of these four statements follows from the hard part, namely, that statement 2 implies statement 3. The equivalence of statements 2 and 6 is immediate from the previous theorem and the Nullstellensatz (or Theorem 2.1). The equivalence of statement 1 and statement 5 is an easy consequence of our work in the next chapter. We verify this equivalence in Chapter 4. Now we prove the most difficult implication, that statement 2 implies statement 3.

Assume statement 2. Then, for any unitary operator, it follows from Theorem 2.1 that $\mathbf{V}(I(U,p)) = \{p\}$. Suppose also that statement 3 fails. Then there is a sequence of unitary matrices $\{U^k\}$ for which

$$\mathbf{D}\left(I\left(U^k, p\right)\right) \to \infty. \tag{142}$$

The coefficients of a unitary operator with respect to any complete orthonormal set are bounded in absolute value by one. Fixing such a complete orthonormal set, we can, after extracting subsequences inductively, assume that the entries $\{U_{ij}^k\}$ are convergent. The resulting operator U^∞ need not be unitary; see Example 4. To get around this point, we also consider the sequence $\{U_{ij}^{k*}\}$ of the entries of the adjoint. For any U^k, we have equality of the ideals

$$I\left(U^k, p\right) = I\left(h, f - U^k g\right) = I\left(h, f - U^k g, U^{k*} f - g\right) \tag{143}$$

because the last components are just linear combinations of the middle ones. The upper semicontinuity of the intersection number guarantees that we always have $\mathbf{D}\left(\text{limit}\left(J^k\right)\right) \geq \text{limit}\left(\mathbf{D}\left(J^k\right)\right)$. This inequality therefore implies that

$$\mathbf{D}\left(h, f - U^\infty g, U^{\infty*} f - g\right) = \infty. \tag{144}$$

This implies, by the Nullstellensatz, that the limiting ideal defines a positive dimensional variety. To verify that this variety must lie in M, we note that the operator norm of the limit is no larger than one. This implies that, for any z in this variety,

$$\|f(z)\| = \|U^\infty g(z)\| \leq \|g(z)\| = \|U^{\infty*} f(z)\| \leq \|f(z)\|. \tag{145}$$

Therefore the chain of inequalities is a chain of equalities, and

$$\|f(z)\| = \|g(z)\|. \tag{146}$$

But this last equality guarantees by Proposition 4 that z lies in some $\mathbf{V}(f - Ug)$. As we are assuming also that $h(z) = 0$, we see that z lies in $\mathbf{V}(I(U,p)) = \{p\}$, so we obtain a contradiction. ∎

Example 4

If the entries of a unitary operator on a Hilbert space converge, then the limiting operator need not be unitary. Consider the following sequence:

$$U^k = \begin{pmatrix} A_k & 0 \\ 0 & Id \end{pmatrix} \tag{147}$$

where the matrix A_k is defined to have entries one on the anti-diagonal and entries zero elsewhere. For any pair of indices i, j there is an integer k_0 such that

$$U_{ij}^k = 0, \quad k > k_0 \tag{148}$$

so the limit matrix is zero. Yet each element in the sequence was unitary. ▯

After we make our results quantitative, we will be able to rephrase a part of Theorem 4 as the following corollary.

COROLLARY 5

(Uniform boundedness principle). *If M is a real analytic real subvariety of $\mathbb{C}^n$, and p lies in M, then precisely one of two alternatives holds:*

1. *There is a complex analytic variety passing through p and lying in M.*

2. *There is a constant c so that every complex analytic variety has order of contact at most c with M at p.*

These ideas apply also to complex varieties of a specific dimension.

THEOREM 5

Let M be a real analytic subvariety of $\mathbb{C}^n$. Let q be a positive integer. Then the following are equivalent:

1. *There is no q-dimensional complex analytic subvariety of $\mathbb{C}^n$ passing through p and lying in M.*

2. *For every unitary U, $\dim (\mathbf{V}(I(U, p))) < q$.*

3. *The point p is a point of finite q-type.*

An important additional result, due to Diederich and Fornaess, appears essentially as a corollary of this theorem in Chapter 4. This result says that the compactness of a real analytic subvariety is sufficient to make every point a point of finite type.

3.3.4   A Bezout theorem in the algebraic case

In this section we apply the quantitative measurements from Chapter 2 to the associated family of ideals $I(U, p)$, in case M is algebraic. For algebraic hypersurfaces this yields an improvement to Theorem 4; it is natural to call the result a Bezout theorem. It says the following: In case p is a point in an algebraic real subvariety M, then either there is a complex analytic subvariety containing this point and lying in M, or there is an explicit bound on the order of contact of such complex varieties. The bound depends on only the dimension and the degree of the defining equation.

THEOREM 6
Suppose that r is a real-valued polynomial of degree d on $\mathbb{C}^n$. Let M denote the zero set of r, and let p lie in M. Then exactly one of two alternatives holds:

1. *There is a complex variety V containing p and lying in M.*

2. *Every complex variety has contact at most $2d(d-1)^{n-1}$ with M at p.*

COROLLARY 6
(Uniform boundedness for points of finite type: algebraic case). *Suppose that $I(U,p) = (h^p, f^p - Ug^p)$ denotes the associated family of holomorphic ideals and that*

$$\forall U, \ \forall p \qquad \mathbf{D}(I(U,p)) < \infty \tag{149}$$

(or, equivalently, M is of finite one type). Then there is a positive integer k so that

$$\mathbf{D}(I(U,p)) < k \qquad \forall U, \ \forall p. \tag{150}$$

PROOF The corollary is an immediate consequence of the proof of the theorem. The proof requires two ingredients. One is the statement that the multiplicity $B(M,p)$ is an upper bound for the maximum order of contact of all complex analytic varieties with M at p. See the inequalities (95) from Chapter 4. The other result is a refinement of the classical Bezout theorem. It is a special case of the refined Bezout theorem of Fulton [Fu] that, if $J = (f_1, \ldots, f_n)$ is any ideal in $_n O$ generated by polynomials, then the codimension (intersection number) satisfies

$$\mathbf{D}(f_1, \ldots f_n) \leq \prod \deg(f_i) \tag{151}$$

when it is finite. Supposing that $I(U,p)$ defines a trivial variety, then a generic choice of n elements generates an ideal I^* that defines also a trivial variety. We then conclude that

$$\mathbf{D}(I(U,p)) \leq \mathbf{D}(I^*) \leq d^n \tag{152}$$

if it is finite, simply because all the generators of the ideal are polynomials of degree at most d. Careful investigation of the definitions of h, f, g in (47)–(49) shows that the degree of each $(f - Ug)_j$ is at most $d-1$ if r is of degree d. The estimate in statement 2 follows. ∎

There is an alternate proof of Corollary 6 in case M is compact (and algebraic). First we fix a point p. If the intersection number $\mathbf{D}(I(U,p))$ is finite for some U_o, then it is finite for nearby U in the usual topology on the unitary group, because this measurement is upper semicontinuous as a function of its parameters. We cover $\mathbf{U}(N)$ by these neighborhoods and can then choose a finite subcollection, because the unitary group is compact. (Note that the compactness is automatic in the algebraic case, but it would hold whenever the number of functions in the decomposition was finite.) The supremum becomes the maximum of a finite set. Hence, for each fixed p, we have a bound $m(p)$ on the intersection number; that is,

$$2 \sup \mathbf{D}(I(U,p)) = B(M,p) < \infty. \tag{153}$$

Now we use the local boundedness of $B(M,p)$ and the compactness of M as above. This shows that the multiplicity and hence also the order of contact are bounded.

From Definitions 6 and 7 we are able to compute numerical invariants. The ideals $I(U,p)$ are more fundamental, however, than the numbers obtained from them. More fundamental yet is the defining equation. If we know the full holomorphic decomposition, then we can reconstruct a defining equation, and hence obtain (in principle) all information. Knowing the ideals does not mean knowing a defining equation; for example, in the strongly pseudoconvex case, all the ideals are the maximal ideal. We cannot distinguish among strongly pseudoconvex hypersurfaces. What we have done is to give a very general class of hypersurface germs (the ones of finite type) whose geometric properties generalize those of strongly pseudoconvex hypersurfaces. The generalization parallels precisely the passage in Chapter 2 from the maximal ideal to ideals primary to it.

4

Points of Finite Type

4.1 Orders of contact

4.1.1 Definition of point of finite type

Now let M be a smooth real hypersurface in $\mathbb{C}^n$. Our aims in this chapter are to make quantitative the ideas of Chapter 3 and extend them to the category of smooth hypersurfaces. In particular we will give several different measurements of the singularity of the Levi form at points where it degenerates, and relate these to each other. We saw in Chapter 1 that there is no proper holomorphic mapping from the polydisc to the ball, precisely because such a mapping would necessarily carry the holomorphic structure in the boundary of the polydisc to the sphere, but the sphere, being strongly pseudoconvex, has no such structure. It is easy to verify that strong pseudoconvexity of a real hypersurface M at a point p is equivalent to the statement that the maximum "order of contact" of complex analytic varieties with M at p is two. The considerations of this section do not rely on pseudoconvexity, but are motivated by trying to generalize the concept of strong pseudoconvexity in this sense. Hence we measure the geometry in terms of algebraic data such as ideals and multiplicities. In case the hypersurface bounds a pseudoconvex domain of finite type, these measurements turn out to be biholomorphic invariants of the domain itself. More generally, we can often decide that there is no proper map between two domains because these boundary invariants do not match appropriately.

Now let us turn to these measurements. We will stratify M into sets where various intersection numbers take different values. Suppose p is a point in M and q is an integer satisfying $1 \leq q \leq n - 1$. In this chapter we define certain numbers

$$B^q(M, p)$$

$$\Delta^q(M, p)$$

$$\Delta^q_{\text{reg}}(M, p) \tag{1}$$

that measure the local geometry at p. These numbers give measurements of the maximum order of contact of q-dimensional ambient complex analytic varieties with M at p. One of the aims of this chapter is to give the precise relationships among these numbers. Our study will emphasize the most important case, when $q = 1$. Recall that C_p^∞ denotes the ring of germs at p of smooth complex-valued functions.

DEFINITION 1 *Let (M, p) be the germ at p of a smooth real hypersurface in $\mathbb{C}^n$. Let r_p denote a generator of the principal ideal in C_p^∞ of functions that vanish on M. The maximum order of contact of complex analytic one-dimensional varieties with M at p is the number*

$$\Delta^1 (M, p) = \mathbf{T} \left(r_p \right) = \sup_z \frac{v \left(z^* r_p \right)}{v(z)} . \tag{2}$$

The point p is called a point of finite 1-type if $\Delta^1 (M, p) < \infty$.

In Chapter 2 we studied the invariant $\mathbf{T}$ when it was applied to ideals in the ring of germs of holomorphic functions. In this chapter we will show how to estimate (from both sides) the value of $\Delta^1 (M, p)$ in terms of $\mathbf{T} (I (U, k, p))$; here $I (U, k, p)$ is a member of the family of ideals in O_p associated with the hypersurface defined by the kth-order Taylor polynomial of some defining function. This reduction amounts to a quantitative version of the results in the algebraic case. When the hypersurface is pseudoconvex, the estimate that relates the invariant $\mathbf{T}$ to $\Delta^1 (M, p)$ is an equality; thus the correspondence is precise under this very natural hypothesis.

So, let $r = r_p$ be a local defining function for (M, p). We denote the kth-order Taylor polynomial of r at p by $j_k r = j_{k,p} r$. This polynomial also defines a real hypersurface in some neighborhood of p. We denote the germ of this hypersurface by (M_k, p). The first observation gives the simple reduction to the algebraic case.

PROPOSITION 1
For a point p on a smooth real hypersurface M, the following statements are equivalent:

1. *The point is a point of finite type, that is, $\Delta^1 (M, p) < \infty$.*
2. *There is an integer k such that $\Delta^1 (M_k, p) \leq k$.*
3. *The sequence of numbers (or plus infinity) $\{\Delta^1 (M_k, p)\}$ is eventually constant and finite.*

PROOF For any smooth function r we can write

$$r = j_k r + (r - j_k r) = j_k r + e_{k+1}. \tag{3}$$

Let z be a parameterized holomorphic curve. We have

$$z^* r = z^* (j_k r) + z^* (e_{k+1}).\tag{4}$$

For any such curve, we must have $v\left(z^* (e_{k+1})\right) \geq (k+1)\, v(z)$. As this is true for each curve, it follows that

$$\Delta^1 (M, p) \leq k \Leftrightarrow \Delta^1 (M_k, p) \leq k.\tag{5}$$

It follows immediately from (5) that statements 1 and 2 are equivalent. It is also clear that if statement 3 holds, then statement 2 does. It remains only to show that statement 2 implies statement 3. It is a consequence of (4) that if $\Delta^1 (M_k, p) \leq k$, then

$$\Delta^1 (N, p) = \Delta^1 (M_k, p)\tag{6}$$

whenever N is any hypersurface defined by any function that has the same k-th order Taylor polynomial at p as $j_k r$. ∎

This proposition deserves reflection. The notion of finite type is (essentially by definition) "finitely determined." A property is finitely determined when it depends on some finite Taylor polynomial of the defining function, but is independent of higher order terms. Example 1 shows that there is a certain amount of subtlety in this.

Example 1
Put

$$r (z, \bar z) = 2 \operatorname{Re} (z_3) + \left| z_1^2 - z_2^3 \right|^2 + \left| z_1^4 \right|^2.\tag{7}$$

Then there is no complex variety lying in this hypersurface. The desired conclusion at the origin follows immediately from Theorem 3.2 because the associated family of holomorphic ideals there consists of one member with finite intersection number:

$$\mathbf{D} \left(z_3, z_1^2 - z_2^3, z_1^4 \right) = 12 < \infty.\tag{8}$$

On the other hand, the surface defined by

$$s (z, \bar z) = 2 \operatorname{Re} (z_3) + \left| z_1^2 - z_2^3 \right|^2 + \left| z_1^4 \right|^2 - \left| z_2^6 \right|^2\tag{9}$$

does contain a complex variety, parameterized by the ubiquitous $\left(t^3, t^2, 0 \right)$, even though the two Taylor polynomials of order 8 are identical. It follows from (5) that any polynomial that agrees with r up to order 12 will also define the origin to be of finite type with $\Delta^1 (M, 0) = 12$. Thus finite type is "finitely determined," but it requires a perhaps higher degree approximation than is first evident. This phenomenon must be taken into account in analytic questions on pseudoconvex domains, but seems to be largely ignored. □

Example 2

The property of pseudoconvexity is not finitely determined, even in the finite type case. Consider a closely related defining equation:

$$s(z, \bar{z}) = 2 \operatorname{Re}(z_3) + \left| z_1^2 - z_2^3 \right|^2 + \left| z_1^4 \right|^2 - \left| z_1 z_2^m \right|^2. \tag{10}$$

Suppose that the integer m is at least six. Then the origin is necessarily a point of finite type, as the last term does not change the type. On the other hand, the resulting hypersurface is not pseudoconvex at the origin for any choice of m. To see this, compute the determinant of the Levi form and evaluate at points p satisfying $z_1(p) = 0$. The result is $-9 |z_2|^{4+2m}$. □

Example 3

It is natural to extend the reasoning in Example 2 to a more general situation. Suppose that

$$r(z, \bar{z}) = 2 \operatorname{Re}(z_n) + \sum_{j=1}^{N} \left| f_j(z') \right|^2 \tag{11}$$

where, as indicated, the functions f_j are independent of z_n. Of course (11) defines a pseudoconvex hypersurface. The determinant of the Levi form is given by

$$\det(\lambda) = \sum \left| J\left(f_{i_1}, \ldots, f_{i_{n-1}}\right) \right|^2, \tag{12}$$

where the sum is taken over all choices of $n-1$ of the functions, and each term is the corresponding Jacobian. Thus the set of weakly pseudoconvex points contains the intersection of the hypersurface with the complex analytic variety defined by these Jacobians. Consider now the effect of changing the defining function by subtracting $|g|^2$ for some holomorphic function g, also independent of z_n. We denote the new Levi form by λ_g. This form will not be, in general, positive semidefinite, as one can see by evaluating the determinant.

$$\det(\lambda_g) = \det(\lambda) - \sum \left| J\left(f_{i_1}, \ldots, f_{i_{n-2}}, g\right) \right|^2. \tag{13}$$

Thus, if g is chosen so that any one of the new terms is not in the radical of the ideal generated by the Jacobians of the $\{f_j\}$, then we violate pseudoconvexity in every neighborhood of the origin. It is obviously possible to do this even when $V(f) = \{0\}$. Thus constructing pseudoconvex examples where the origin is of finite type, but where the pseudoconvexity is not finitely determined, as in Example 2, is quite easy. □

Elaboration on the examples

A strongly pseudoconvex point p on a real hypersurface M satisfies two nice properties. First, any hypersurface M' containing p that agrees with M up to second order must be also strongly pseudoconvex. Second, as a consequence, the maximum order of contact at p of ambient complex analytic varieties with M' remains two also. For hypersurfaces of finite type, we obtain an analog of the second property. There is no analog of the first property; Examples 2 and 3 reveal that a pseudoconvex hypersurface of finite type can be perturbed so that the maximum order of contact is unaltered, yet such that the pseudoconvexity is altered. If we replace the term $-|z_1 z_2^m|^2$ in Example 2 with $-|z_1|^2 \phi(z_2, \overline{z_2})$, where ϕ vanishes to infinite order at zero and is otherwise positive, then we see that pseudoconvexity cannot be determined by the Taylor series at a point.

4.1.2 Inequalities on the order of contact

Proposition 1 shows that one can determine whether a point in a smooth hypersurface is a point of finite type by examining the functions $\Delta^1(M_k, p)$. Each Taylor polynomial $j_{k,p} r$ has a holomorphic decomposition

$$j_{k,p} r = 2 \operatorname{Re}\left(h_{k,p}\right) + \|f_{k,p}\|^2 - \|g_{k,p}\|^2 \tag{14}$$

and thereby defines associated families of ideals $I(U, k, p) = \left(h_{p,k}, f_{p,k} - U g_{p,k}\right)$. For simplicity of notation we fix the integer k and the point p, and drop the subscripts on the functions. The following theorem gives the relationship between the orders of contact of ideals in the two different rings. Later (Theorem 7) we improve (15) to an equality, under the additional assumption that the hypersurface is pseudoconvex.

THEOREM 1
The maximum order of contact of parameterized holomorphic curves with the algebraic real hypersurface M_k at the point p satisfies the double inequality

$$\sup_U \mathbf{T}\left(I(U, p, k)\right) + 1 \le \Delta^1(M_k, p) \le 2 \sup_U \mathbf{T}\left(I(U, p, k)\right). \tag{15}$$

PROOF To fix notation suppose that M_k is defined by the vanishing of

$$\Phi = 2 \operatorname{Re}(h) + \|f\|^2 - \|g\|^2. \tag{16}$$

We may assume that the point p is the origin. When the right side of (15) is infinite, Theorem 2 of Chapter 3 guarantees that $\Delta^1(M_k, p)$ is infinite. In this case there may be a unique parameterized holomorphic curve lying in M_k. When the right side of (15) is finite, there are many parameterized holomorphic curves with maximal order of contact. Consider such a curve z. Supposing that

$$\Delta^1(M_k, p, z) = \frac{v(z^* \Phi)}{v(z)} \tag{17}$$

is maximal, we claim that we can find another curve ζ such that $\zeta^* h = 0$ and

$$\Delta^1 (M_k, p, \zeta) = \frac{v (\zeta^* \Phi)}{v (\zeta)} \tag{18}$$

is still maximal. To prove the claim, notice that since M_k is a hypersurface, the function h vanishes to first order, and we may assume therefore that it is the coordinate function z_n. We write the curve as $z = (z', z_n)$. One can rewrite the defining function as

$$\Phi = 2 \operatorname{Re} (z_n) + z_n \overline{a (z, \bar{z})} + \overline{z_n} a (z, \bar{z}) + B (z', \overline{z'}) \tag{19}$$

where the last term is independent of z_n, and where $a (0) = 0$. Consider the pullback

$$z^* \Phi = 2 \operatorname{Re} (z_n (t)) + A (t, \bar{t}) + B (z' (t), \overline{z' (t)}) . \tag{20}$$

The second term in (20) vanishes to strictly higher order than the first, so it can be ignored. If the third term vanishes to lower order than the first, then it also follows that $v (z) < v (z_n)$. This implies that the curve ζ, where

$$\zeta (t) = (z_1 (t), \ldots, z_{n-1} (t), 0) \tag{21}$$

satisfies

$$\frac{v (\zeta^* \Phi)}{v (\zeta)} = \frac{v (z^* \Phi)}{v (z)} \tag{22}$$

as both numerators and both denominators are equal. This is what we want to show. We claim that the other case does not occur for a curve for which $\Delta^1 (M_k, p, z)$ is maximal. To see this, consider again two cases. If $v (z) = v (z_n)$, then $v (z^* \Phi) / v (z)$ equals one; this isn't maximal as the other term vanishes to order at least two along every curve. If, in the other case $v (z) = v (z')$, then replacing z by ζ increases the numerator on the left side in (22) without changing the denominator. This again contradicts the maximality. Thus the claim is proved. In more geometric terms, the claim states that, of the curves with maximal order of contact, at least one must lie in the holomorphic tangent space.

Now let ζ be such a parameterized holomorphic curve. Thus we assume that $\zeta^* h = 0$ and

$$\Delta^1 (M_k, p, \zeta) = \Delta^1 (M_k, p) = \frac{v (\zeta^* \Phi)}{v (\zeta)} . \tag{23}$$

Fix a unitary matrix U. For purposes of this proof, rewrite the expression

$$\begin{aligned}
\zeta^* \Phi &= \|\zeta^* f\|^2 - \|\zeta^* g\|^2 \\
&= \|\zeta^* (f - Ug)\|^2 + 2 \operatorname{Re} \langle \zeta^* (f - Ug), \zeta^* (Ug) \rangle .
\end{aligned} \tag{24}$$

Notice that $(Ug)(0) = 0$. From this it follows that $v(\zeta^*(Ug)) \geq v(\zeta)$. Combining this with (23) reveals that

$$\Delta^1(M_k, p) = \frac{v(\zeta^*\Phi)}{v(\zeta)}$$

$$\geq \min\left(\frac{2v(\zeta^*(f - Ug))}{v(\zeta)}, \frac{v(\zeta^*(f - Ug)) + v(\zeta))}{v(\zeta)}\right)$$

$$\geq \frac{v(\zeta^*(f - Ug))}{v(\zeta)} + 1. \tag{25}$$

This formula holds for an arbitrary unitary matrix. By taking the supremum over the unitary matrices, we obtain the first desired inequality. It remains to prove the second inequality. We use the same notation as in the first part. Again we may assume that $\zeta^*h = 0$. Choose a unitary matrix U. There are vectors L_1, L_2 and integers m, s so that we can write

$$\zeta^*(f - Ug) = L_1 t^m + \cdots$$

$$\zeta^*(Ug) = L_2 t^s + \cdots. \tag{26}$$

Computation shows then that

$$\zeta^*\Phi = \|L_1\|^2 |t|^{2m} + 2 \operatorname{Re}\langle L_1, L_2\rangle t^m \bar{t}^s + \cdots. \tag{27}$$

When $m < s$ we get the desired conclusion, and equality is attained for the right-hand side. When $m > s$, the order of vanishing is $m + s$, which again implies what we want. The problem case is if they are equal and cancellation occurs. We claim that if this happens, then the unitary matrix must not have been chosen so as to give the maximum value in the definition of the invariant $\mathbf{T}(I(U, k, p))$. The only way cancellation can occur is if

$$L_2 = -\frac{1}{2}L_1 + w \quad \text{where} \quad 2 \operatorname{Re}\langle L_1, w\rangle = 0. \tag{28}$$

In this case, we add the two equations in (26) to see that

$$\zeta^*f = \left(\frac{1}{2}L_1 + w\right) t^m + \cdots = A t^m + \cdots$$

$$\zeta^*g = \left(-\frac{1}{2}U^{-1}L_1 + U^{-1}w\right) t^m + \cdots = B t^m + \cdots. \tag{29}$$

The vectors A, B have the same length. To verify this, note that

$$\|A\|^2 = \|\frac{1}{2}L_1 + w\|^2$$

$$= \|-\frac{1}{2}U^{-1}L_1 + U^{-1}w\|^2$$

$$= \|B\|^2 \tag{30}$$

since U is unitary and also because the cross term $2 \operatorname{Re} \langle \frac{1}{2}L_1, w \rangle = 0$ vanishes. Therefore there is another unitary matrix satisfying

$$A = U^{\#}B. \tag{31}$$

This means that

$$\zeta^* \left(f - U^{\#}g\right) = \left(A - U^{\#}B\right) t^m + \cdots = L\, t^{m+1} + \cdots \tag{32}$$

and therefore implies that $v\left(\zeta^*\left(f - U^{\#}g\right)\right) > v\left(\zeta^*\left(f - Ug\right)\right)$. The conclusion is that the only way cancellation can take place is if we didn't pick the unitary matrix for which $\mathbf{T}\left(I\left(U, k, p\right)\right)$ is maximal. Thus, by taking the supremum over the unitary matrices, the result follows, finishing the proof. ∎

This pair of inequalities has many consequences, so this theorem is one of the important technical results of this chapter. The heuristic idea behind the inequality is that the information given by the associated family of ideals is sufficient to estimate from both sides the order of contact of complex analytic varieties with a hypersurface. In many cases there is an equality. This is easy to show when the domain has a defining function that can be written

$$r = 2 \operatorname{Re}(h) + \|f\|^2 . \tag{33}$$

For many considerations this class of domains exhibits simplified behavior, yet it is sufficiently general that many of the analytic difficulties occur already within this class. We discuss these domains further in the section on the real analytic case.

4.1.3　Openness of the set of points of finite type

It is time to prove that the set of points of finite type on a real hypersurface M is an open subset of M. This is also true for a real subvariety. The proof will also enable us to prove an analogous result for families of hypersurfaces. Thus "finite type" is a nondegeneracy condition. We will give analogous proofs also for points of finite q-type also. For now let us suppose that q equals one.

THEOREM 2

Let M be a real hypersurface of $\mathbb{C}^n$. The set of points of finite type is then an open subset of M. The type $p \to \Delta^1(M, p)$ is a locally bounded function on the hypersurface.

PROOF We produce a chain of inequalities that give a local bound for the type. Suppose that p_o is a point of finite type, and $j_{k,p}r$ denotes the kth-order Taylor polynomial of some defining function. Whether or not p is a point of finite type, there are arbitrarily large integers k such that the first line of (34) holds. The second line follows because of Theorem 2.2, and the third line by the upper semicontinuity (see Theorem 2.1) of the codimension operator **D**. The fourth line holds by inequality (136) of Chapter 2, and the last by inequality (15) from Theorem 1. Thus, when p_o is a point of finite type, these inequalities show that a nearby point p is also:

$$\begin{aligned}
\Delta^1(M, p) &\leq 2 \sup_U \mathbf{T}(I(U, k, p)) \\
&\leq 2 \sup_U \mathbf{D}(I(U, k, p)) \\
&\leq 2 \sup_U \mathbf{D}(I(U, k, p_o)) \\
&\leq 2 \sup_U \left(\mathbf{T}(I(U, k, p_o))^{n-1} \right) \\
&\leq 2 \left(\Delta^1(M, p_o) - 1 \right)^{n-1}.
\end{aligned} \tag{34}$$

Notice that we have used the fact that, on a hypersurface, there is always at least one function with nonvanishing derivative, namely h, in the associated family of ideals. After a change of coordinates we may assume that this function is linear, and apply the inequality (136) of Theorem 2.2 with $q = 1$. This finishes the proof for a hypersurface. ∎

COROLLARY 1

If M is a compact smooth real hypersurface, and each point on M is a point of finite type, then the type $\Delta^1(M, p)$ is uniformly bounded on M.

PROOF A locally bounded function on a compact set is globally bounded. ∎

It is easy to see that compactness is required in Corollary 1. We make a few remarks about the case when M is a subvariety. This situation is only slightly different. We must first replace $n - 1$ by n in (34) because we may not have any functions with nonvanishing derivative in the ideal. Second we must replace the far left-hand side in the double estimate (15) in Theorem 1 for the maximum order of contact by $\sup_U \mathbf{T}(I(U, p, k))$. We leave it to the reader to verify these points. We therefore only state the result in the subvariety case.

The proof of Theorem 2 is complete in the hypersurface case; the author knows of no application of the more general Theorem 3.

THEOREM 3
Let M be a real subvariety of $\mathbb{C}^n$. The set of points on M of finite one-type is an open subset of M.

The same conclusion holds for points of finite q-type for each integer q. This will be proved in Theorem 6.

4.1.4 More on the real analytic case

Suppose that r_p is real analytic near p, that

$$r_p(z, \bar{z}) = 2 \operatorname{Re}\left(h_p(z)\right) + \sum_{|a|,|b|=1}^{\infty} c_{ab}(p)(z-p)^a \left(\overline{z-p}\right)^b, \tag{35}$$

and that the matrix of coefficients $c_{ab}(p)$ is positive semidefinite. This condition is necessary and sufficient for writing the holomorphic decomposition without any g's. Thus we can write

$$r_p(z, \bar{z}) = 2 \operatorname{Re}(h(z)) + \|f(z)\|^2. \tag{36}$$

Now it follows from Lemma 2.5 that

$$\Delta^1(M, p) = 2 \, \mathbf{T}(h, f_1, f_2, \ldots) = 2 \, \mathbf{T}(h, f). \tag{37}$$

Since the ring of germs of holomorphic functions at p is Noetherian, we can choose a finite subset of the $\{f_j\}$ that together with h generate the ideal, thus making things even simpler. We have an easy version of the basic equivalence in this case, as statements 4 and 5 in the important Theorem 4 below are obviously then equivalent.

Lempert [Le1] has proved, using nontrivial methods from the theory of the Szegö kernel function, that every bounded real analytic strongly pseudoconvex domain can be defined by an equation of the form $r(z, \bar{z}) = \|f(z)\|^2 - 1$. Here f is in general a countable collection of functions, but the squared norm is convergent. We have been writing the defining equation by isolating the pure term h. A defining equation of the form

$$4 \operatorname{Re}(h) + \|f\|^2 \tag{38}$$

can be transformed into an equation of the form

$$\|f^{\#}\|^2 - 1 \tag{39}$$

by writing $4 \operatorname{Re}(h) = |h+1|^2 - |h-1|^2$ and multiplying by the (local) unit $|h-1|^{-2}$. The class of domains with a positive definite defining form—that is, those for which there is a defining function without any negative squares—is a subclass for which many considerations in the general theory are easier. In particular, it is easy to verify that the function h is a holomorphic support function. See [Ra] for some uses of such a function. For the purposes of determining orders of contact of complex analytic varieties, the advantage such domains offer is that there are no unitary matrices involved. The general case is more complicated because one must study a family of ideals rather than just one.

Before turning to the main theorem about finite one-type in the real analytic case, we make several other comments about Lempert's work. (See the survey [Fo3] for more details on the remarks in this paragraph.) Work of Faran and Forstneric showed that it is impossible to find a proper holomorphic mapping from a bounded strongly pseudoconvex domain with real analytic boundary to a ball in some $\mathbb{C}^N$ that is smooth on the closure. Thus it is not possible to choose only finitely many f_j. Such embeddings do exist if we eliminate the smoothness requirement at the boundary. Lempert's result shows that it is possible to find a proper embedding if we allow the ball to lie in an (infinite dimensional) Hilbert space. Furthermore, the functions are holomorphic in a neighborhood of the closed domain and give a global defining function. It is natural to ask whether there is a global defining function of the form $2 \operatorname{Re}(h) + ||f||^2 - ||g||^2$ for general real analytic domains. In the same paper [Le1], Lempert shows that there are bounded pseudoconvex domains with real analytic boundary (thus of finite type by the Diederich–Fornaess theorem) that do not have such a global defining function.

The following theorem summarizes what we have done so far in the real analytic case. In the proof we finish those points left unresolved in the proof of Theorem 3.4.

THEOREM 4

Suppose M is a real analytic real subvariety of $\mathbb{C}^n$ and p lies in M. Let $I(U,p)$ be the associated family of holomorphic ideals. The following are equivalent:

1. *The point p is a point of finite one-type.*

2. *For every unitary U, the ideal $(h, f - Ug) = I(U,p)$ is primary to the maximal ideal; that is,*

$$\operatorname{rad}(I(U,p)) = \mathbf{M}. \tag{40}$$

3. *The intersection number (codimension) of each ideal is finite. That is, for every unitary U,*

$$\mathbf{D}(I(U,p)) < \infty. \tag{41}$$

4. *The order of contact of every ideal is finite. That is, for every unitary U,*

$$\mathbf{T}(I(U,p)) < \infty. \tag{42}$$

5. *The intersection numbers of the ideals $I(U,p)$ are uniformly bounded. That is, there is a positive integer k such that*

$$\mathbf{D}(I(U,p)) < k \quad \forall U. \tag{43}$$

6. *The orders of contact of the ideals $I(U,p)$ are uniformly bounded. That is, there is a positive constant Δ such that*

$$\mathbf{T}(I(U,p)) < \Delta \quad \forall U. \tag{44}$$

7. *There is no positive dimensional germ of a complex analytic subvariety of $\mathbb{C}^n$ that contains p and lies in M.*

PROOF We have proved already in Theorem 3.4 that the statements 3 through 7 are equivalent. Statement 2 is equivalent to statement 3, by the Nullstellensatz; see also the discussion after Definition 2.4. To verify that statement 1 is equivalent to all these, we observe that the definition of finite type means that there is a constant Δ^1 so that

$$\sup_z \frac{v(z^*r)}{v(z)} \leq \Delta^1 \tag{45}$$

and hence finite type implies statement 7. On the other hand, suppose that the point is not a point of finite type. We wish to contradict statement 4. To do so, we need the analog of the inequalities (15) for the real analytic case. That is, we claim

$$\sup_U \mathbf{T}(I(U,p)) + 1 \leq \Delta^1(M,p) \leq 2 \sup_U \mathbf{T}(I(U,p)) \tag{46}$$

holds in the real analytic case. The proof of (46) is virtually the same as the proof of Theorem 1 and will be omitted. This gives the necessary contradiction, and finishes the proof of the theorem. ∎

It is a well known theorem of Diederich and Fornaess that a compact real analytic subvariety can contain no positive dimensional complex analytic varieties; combining their theorem with Theorem 4 enables us to conclude that every point on a compact real analytic subvariety is a point of finite type. Let us give now a simple proof of the Diederich–Fornaess theorem based on the associated ideals of holomorphic functions.

THEOREM 5
(Diederich–Fornaess). *A compact real analytic subvariety of $\mathbb{C}^n$ contains no positive dimensional complex analytic varieties.*

PROOF Let M be the given real analytic subvariety. Let W denote the set of points p in M for which there is a complex analytic variety of positive dimension passing through p and lying in M. The complement of W consists of points of finite one-type by Theorem 4. From Theorem 1 we know that W is a closed subset of M, so the compactness of M implies the compactness of W. Suppose that W is nonempty. Choose a point of W at greatest distance from the origin. The holomorphic function

$$w(z) = \frac{1}{1 - \langle z, p \rangle + ||p||^2} \tag{47}$$

achieves its maximum on W at p. According to Theorem 3.3 there is a unitary matrix such that the variety $\mathbf{V}(I(U, p))$ is positive dimensional and lies in M. The point is in its interior, yet the function $w \mid \mathbf{V}(I(U, p))$ achieves its maximum there. Since this function is not constant, the maximum principle is violated. This contradiction means that M cannot be compact unless W is empty. This is precisely the conclusion of the theorem. ∎

4.2 Local bounds

4.2.1 The pseudoconvex case

Our aim now is to give improvements on the bounds that arise in the proof of openness. These improvements arise if we assume that the hypersurface is pseudoconvex, or if we assume that there are a certain number of positive eigenvalues of the Levi form. There is a general belief that these sharp bounds should play a significant role in the study of analytic objects on pseudoconvex domains, but the subject is not yet sufficiently developed to see this.

We discuss first a nice example where the maximum jump occurs.

Example 4

Consider the algebraic real hypersurface defined by

$$r(z, \bar{z}) = 2 \operatorname{Re}(z_n) + \sum_{j=1}^{n-2} \left| z_j^d - z_{j+1} z_n \right|^2 + \left| z_{n-1}^d \right|^2. \tag{48}$$

The associated family of holomorphic ideals at the origin is the single ideal

$$\left(z_1^d - z_n z_2, z_2^d - z_n z_3, \ldots, z_{n-2}^d - z_n z_{n-1}, z_{n-1}^d, z_n \right)$$
$$= \left(z_1^d, \ldots, z_{n-1}^d, z_n \right). \tag{49}$$

We have

$$\Delta^1(M, 0) = 2\mathbf{T}\left(z_1^d, \ldots, z_{n-1}^d, z_n \right) = 2d. \tag{50}$$

Let us choose a nearby point of the form $(0, 0, \ldots, 0, i\epsilon)$ and consider the holomorphic curve defined by

$$\left(t, \frac{t^d}{i\epsilon}, \frac{t^{d^2}}{(i\epsilon)^{d+1}}, \ldots, \frac{t^{d^{n-2}}}{(i\epsilon)^{\frac{1-d^{n-2}}{1-d}}}, i\epsilon \right). \tag{51}$$

We substitute (51) into the defining equation. This yields

$$z^* r = \frac{|t|^{2d^{n-1}}}{c(\epsilon)} \tag{52}$$

and therefore $v(z^* r) = 2d^{n-1}$. Since the curve has multiplicity equal to one, we see that the order of contact nearby can be at least as large as $2d^{n-1}$. This is of course greater than $2d$, the maximum order of contact at the origin. This is the worst possible jump in value for a pseudoconvex hypersurface; Theorem 6 shows that it is sharp for pseudoconvex hypersurfaces. For pseudoconvex subvarieties one can get one higher power, and for general subvarieties, the sharp bound is a power of two larger. After some more discussion of this example, we investigate the sharpness of the bounds.   ☐

Continuation of Example 4

Take $n = 3$, $d = 2$ in the example, to obtain the defining function

$$r(z, \bar{z}) = 2 \operatorname{Re}(z_3) + \left| z_1^2 - z_2 z_3 \right|^2 + \left| z_2^2 \right|^2. \tag{53}$$

Then we have the values

$$\Delta^1(M, 0) = 4$$

$$\Delta^1(M, p) = 8 = 2 \left(\frac{4}{2} \right)^{3-1} \qquad p = (0, 0, ia). \tag{54}$$

This defining equation gives perhaps the simplest possible example where the order of contact fails to be upper semicontinuous. It is interesting that this phenomenon can occur for domains defined by polynomials of degree four. The failure of upper semicontinuity does not depend on singular curves.

There is a sharp bound in the pseudoconvex case. We have the following estimate for the type.

THEOREM 6

Suppose M is a pseudoconvex real hypersurface in $\mathbb{C}^n$, and the Levi form at a point p_o has q positive eigenvalues. Then there is a neighborhood of p_o on which

$$\Delta^1(M,p) \leq \frac{\left(\Delta^1(M,p_o)\right)^{n-1-q}}{2^{n-q}}. \tag{55}$$

PROOF Recall the basic string of inequalities

$$
\begin{aligned}
\Delta^1(M,p) &\leq 2 \sup_U \mathbf{T}\left(I(U,k,p)\right) \\
&\leq 2 \sup_U \mathbf{D}\left(I(U,k,p)\right) \\
&\leq 2 \sup_U \mathbf{D}\left(I(U,k,p_o)\right) \\
&\leq 2 \sup_U \left(\mathbf{T}\left(I(U,k,p_o)\right)^n\right) \\
&\leq 2 \left(\Delta^1(M,p_o) - 1\right)^n. \tag{56}
\end{aligned}
$$

The improvements arise as follows. First, if M is a hypersurface, then there is at least one function with nonvanishing differential in the ideal $I(U,k,p_o)$. For each positive eigenvalue of the Levi form, we get an additional such function. This follows from the definitions (48) and (49) from Chapter 3 of the functions in the holomorphic decomposition. According to the implicit function theorem, we may choose coordinates so that these functions are independent linear functions. If the Levi form on a hypersurface has precisely q positive eigenvalues, then we may assume that the ideal contains $q + 1$ independent linear functions. Theorem 2.4 gives then the estimate

$$\mathbf{D}\left(I(U,k,p_o)\right) \leq \left(\mathbf{T}\left(I(U,k,p_o)\right)\right)^{n-1-q}. \tag{57}$$

The other improvement is that, in the pseudoconvex case, we claim

$$\Delta^1(M,p_o) = 2 \sup_U \left(\mathbf{T}\left(I(U,k,p_o)\right)\right) \tag{58}$$

for k sufficiently large. Replacing the penultimate inequality in (56) by (57) and the last inequality in (56) by (58) introduces a factor of $1/2^{n-1-q}$ and implies the result. ∎

Thus it is necessary only to verify (58). We state this below as Theorem 7. In order to prove it, we need the following lemma which shows the simplest constraint that subharmonicity puts on the Taylor series of a smooth function. Determination of all the conditions for the positivity of a trigonometric polynomial is rather intricate, but not needed here.

LEMMA 1

Suppose that $u\left(t,\bar{t}\right) = \sum_{a+b=m} c_{ab} t^a \bar{t}^b +$ higher order terms *is a smooth subharmonic function in the plane. Assume as indicated that* $u(0) = 0$, u *vanishes to finite order* m *there, and it contains no pure terms. Then the order of vanishing is even, that is,* $m = 2k$, *and* $c_{kk} > 0$.

PROOF   We are assuming that

$$\frac{\partial^2 u}{\partial t \partial \bar{t}} \geq 0. \tag{59}$$

Setting $t = |t| e^{i\theta}$, and choosing $|t|$ sufficiently small, we see that

$$\frac{\partial^2}{\partial t \partial \bar{t}} \left(\sum_{a+b=m} c_{ab} t^a \bar{t}^b \right) \geq 0 \tag{60}$$

also. Since there are no pure terms, the trigonometric polynomial is not harmonic. Therefore the average value of (60) over a circle is positive. Writing the average as an integral we have

$$\frac{1}{2\pi} \int_0^{2\pi} \frac{\partial^2}{\partial t \partial \bar{t}} \left(\sum_{a+b=m} c_{ab} t^a \bar{t}^b \right) d\theta > 0. \tag{61}$$

On the other hand, all terms in the integrand integrate to zero except those for which $a = b$, so we conclude that $a^2 c_{aa} > 0$, which is what we wanted to show. ∎

PROPOSITION 2

Suppose that M *is a pseudoconvex hypersurface containing the origin with local defining function* r. *Suppose further that* $z : (\mathbb{C}, 0) \rightarrow (\mathbb{C}^n, 0)$ *is a parameterized holomorphic curve such that the Taylor series for* $z^* r$ *satisfies*

$$v\left(z^* r\right) = m$$

$$\left(\frac{\partial}{\partial t}\right)^a \left(z^* r\right)(0) = 0 \qquad a \leq m. \tag{62}$$

Then the order of vanishing $m = 2k$ *is even, and the coefficient of* $|t|^{2k}$ *in* $z^* r$ *is positive.*

PROOF   We may suppose without loss of generality that the defining function in a neighborhood of the origin can be written

$$r(z, \bar{z}) = 2 \operatorname{Re}(z_n) + f\left(z', \overline{z'}\right) + 2 \operatorname{Im}(z_n) g\left(z', \overline{z'}, 2 \operatorname{Im}(z_n)\right) \tag{63}$$

for appropriate smooth functions f, g. Let us write $z_n = x + iy$ and write g' and g'' for the first and second derivatives of g with respect to the last variable. The pseudoconvexity of M is equivalent to the semidefiniteness of the matrix

$$\lambda_{i\bar{j}} = \frac{r_{i\bar{j}}r_n r_{\bar{n}} - r_{i\bar{n}}r_n r_{\bar{j}} - r_{n\bar{j}}r_i r_{\bar{n}} + r_{n\bar{n}}r_i r_{\bar{j}}}{|r_n|^2} \tag{64}$$

according to Proposition 3.1. Expressing this matrix in terms of the formula (63) for the defining function, we see that the following matrix is semidefinite at points of the hypersurface:

$$
\begin{aligned}
\left(f_{k\bar{j}} + 2yg_{k\bar{j}}\right) &\left|1 - ig - 2iyg'\right|^2 + \left(2yg'' + 2g'\right)\left(f_k + 2yg_k\right)\left(f_{\bar{j}} + 2yg_{\bar{j}}\right) \\
&+ \left(ig_k + 2iyg_k'\right)\left(2iyg' + ig\right)\left(f_{\bar{j}} + 2yg_{\bar{j}}\right) \\
&+ \left(ig_{\bar{j}} + 2iyg_{\bar{j}}'\right)\left(2iyg' + ig\right)\left(f_k + 2yg_k\right). \quad (65)
\end{aligned}
$$

In (65), we can let $y \to 0$ because we can then adjust x so that we stay on the hypersurface. This shows that the matrix

$$\left(f_{k\bar{j}}\right)|1 - ig|^2 + f_k f_{\bar{j}} 2g' - gg_k f_{\bar{j}} - gf_k g_{\bar{j}} \tag{66}$$

is semidefinite at all nearby points of the form $\left(z', \overline{z'}, 0\right)$. Now we consider the function $t \to r(z(t))$ where we assume that $z_n(t) = 0$. Its Laplacian satisfies

$$\frac{\partial}{\partial t}\frac{\partial}{\partial \bar{t}}(r(z(t))) = \sum_{k,j=1}^{n-1} f_{k\bar{j}}\xi_k\overline{\xi_j} \tag{67}$$

where $\xi = z'(t)$. We claim that (66) implies that (67) is nonnegative. Supposing that f vanishes to order m at the origin, we see that the first term in (66) is of order $m - 2$, while the others are of order at least $2m - 2$, $m - 1$, $m - 1$, respectively. Dividing by $\|z'\|^{m-2}$ and letting z' tend to zero shows that f is plurisubharmonic near the origin and hence implies the claim. The claim together with Lemma 1 then proves Proposition 2. ∎

To finish the proof of Theorem 6, we must verify (58). The following theorem does so; its conclusion nicely combines the ideas from Chapters 2 and 3 when the hypersurface is pseudoconvex.

THEOREM 7

Suppose M is a pseudoconvex real hypersurface in $\mathbb{C}^n$ with local defining function r near a point p of finite type. Let $I(U, k, p)$ denote the family of holomorphic ideals associated with the kth-order Taylor polynomial of this defining function. Then the maximum order of contact of one-dimensional complex an-

alytic varieties with M at p satisfies $\Delta^1(M,p) = 2 \sup_U \mathbf{T}(I(U,k,p))$ for all sufficiently large k.

PROOF Since p is a point of finite type, we may choose a sufficiently large integer k so that it suffices to work with the algebraic hypersurface germ (M_k, p). We then repeat the proof of the second inequality (15) from Theorem 1, up to (27). If $m < s$ in (27), then we conclude that $\Delta^1(M,p) = \Delta^1(M_k,p) = 2 \sup_U \mathbf{T}(I(U,k,p))$, which is the desired conclusion. As in that proof, we cannot have cancellation when $m = s$. The remaining possibility is when $m > s$, and the lowest order term in (27) arises from the second term in (24). By pseudoconvexity and Proposition 2, the order of vanishing must be even, say $2d$, and the coefficient of $|t|^{2d}$ is positive. By (24) and (26), $m \leq d$, $s \leq d$. Using the inequality $m > s$, we see that $2d \geq m + s > 2s$, which is a contradiction. In the pseudoconvex case, therefore, the lowest order term in (24) must arise from $\|\zeta^*(f - Ug)\|^2$. This finishes the proof of Theorem 7, and hence also finishes the proof of Theorem 6. ∎

4.2.2 Families of hypersurfaces

The proof of openness applies almost immediately to the situation when one has a family of hypersurfaces passing through the same point and depending continuously on a parameter. Thus we assume that $M(\epsilon)$ is a real hypersurface with defining equation $r(z, \bar{z}, \epsilon)$, where the Taylor coefficients of r are all continuous functions in ϵ. We suppose that the origin lies in $M(\epsilon)$ for all ϵ in a neighborhood of ϵ_o. Then we have the following result concerning dependence on parameters.

THEOREM 8
The function

$$\epsilon \to \Delta^1(M(\epsilon), 0) \tag{68}$$

is locally bounded in ϵ. Furthermore the following inequalities are valid for ϵ near ϵ_o:

1. $\Delta^1(M(\epsilon), 0) \leq 2\left(\Delta^1(M(\epsilon_o), 0) - 1\right)^{n-1}$

2. $\Delta^1(M(\epsilon), 0) \leq 2\left(\dfrac{\Delta^1(M(\epsilon_o), 0)}{2}\right)^{n-1-q}$. $\tag{69}$

The second holds if all the hypersurfaces are pseudoconvex and if the Levi form on the hypersurface $M(\epsilon_o)$ has q positive eigenvalues.

PROOF The proof is just a repetition of the string of inequalities (34) used to prove openness. Choose a large integer k. Including the dependence on this parameter in the notation, we have the same chain of inequalities as before.

$$\Delta^1 \left(M\left(\epsilon\right)_k, 0 \right) \leq 2 \sup_U \mathbf{T}\left(I\left(\epsilon, U, k, 0\right) \right)$$

$$\leq 2 \sup_U \mathbf{D}\left(I\left(\epsilon, U, k, 0\right) \right)$$

$$\leq 2 \sup_U \mathbf{D}\left(I\left(\epsilon_o, U, k, 0\right) \right)$$

$$\leq 2 \sup_U \left(\mathbf{T}\left(I\left(\epsilon_o, U, k, 0\right) \right)^{n-1} \right)$$

$$\leq 2 \left(\Delta^1 \left(M\left(\epsilon_o\right)_k, 0 \right) - 1 \right)^{n-1}. \tag{70}$$

By hypothesis, the origin is a point of finite type on the hypersurface $M\left(\epsilon_o\right)$, so the last collection of numbers is eventually constant. The first conclusion now follows from Proposition 1. The second statement is also a repetition of the proof of the sharp bound in the pseudoconvex case, and can be left to the reader. ∎

4.3 Other finite type conditions

4.3.1 Finite type in two dimensions

The original definition of point of finite type was restricted to domains in two dimensions. Kohn was seeking conditions for subelliptic estimates on pseudoconvex domains, and was led to the study of iterated commutators. In this section we show that this condition and several others are all equivalent, in the special two-dimensional case. We make some of the definitions in general, even though the theorem does not hold in higher dimensions.

DEFINITION 2   Suppose that M is a CR manifold, and L is a type (1,0) vector field defined near a point p in M. We say that the type of L at p is k, and write

$$\text{type}_p L = k \tag{71}$$

if the following holds. There is an iterated commutator

$$\left[\ldots \left[\left[L_1, L_2\right], L_3\right], \ldots, L_k \right] = \Psi_k \tag{72}$$

where each L_j is either L or $\overline{L}$, such that

$$\langle \Psi_k, \eta \rangle \left(p\right) \neq 0 \tag{73}$$

and k is the smallest such integer.

DEFINITION 3   *We say that $c_p(L)$ equals k, if k is the smallest integer for which there is a monomial $D^{k-2} = \prod_{j=1}^{k-2} L_j$ for which*

$$D^{k-2} \lambda\left(L, \overline{L}\right)(p) \neq 0. \tag{74}$$

Again each L_j is either L or $\overline{L}$.

In case M is a real hypersurface with defining function r, we write $r \mid V$ for the restriction of r to the complex hyperplane tangent to M.

THEOREM 9
Suppose M is a real hypersurface in $\mathbb{C}^2$, and let L be a nonzero $(1,0)$ vector field defined near p. The following numbers are all equal:

1. $\Delta^1(M, p)$
2. $\Delta^1_{\text{reg}}(M, p)$
3. $\text{type}_p L$
4. $c_p(L)$
5. $v_p\left(r \mid V\right)$.

PROOF   The first thing we do is give an intrinsic proof of the equality of the numbers $\text{type}_p L$ and $c_p(L)$ in case M is a three-dimensional CR manifold. Suppose that we make any choice of a vector field T not in $T^{1,0}(M) \oplus T^{0,1}(M)$ such that $T = -\overline{T}$ and such that T does not vanish at p. According to the formulas (see Proposition 3.1) in Chapter 3, we have the function α_L defined by

$$\langle [T, L], \eta \rangle = \alpha_L. \tag{75}$$

This number expedites writing the iterated commutators. Also, $\alpha_{\overline{L}} = \overline{\alpha_L}$. We begin an induction. Notice that the numbers $\text{type}_p L$ and $c_p(L)$ are simultaneously equal to two by definition. Suppose that we have an iterated commutator of the form

$$\Upsilon_{k+3} = \left[\ldots \left[\left[\left[L, \overline{L}\right], L_1\right], L_2\right], \ldots, L_{k+1}\right] \tag{76}$$

where each L_j is either L or $\overline{L}$. We suppose inductively that there is a differential operator P_j of order j, for each $j \leq k$, such that

$$\langle \Upsilon_{k+2}, \eta \rangle = \prod_{j=1}^{k} \left(\alpha_{L_j} - L_j\right) \lambda\left(L, \overline{L}\right) + P_{k-1} \lambda\left(L, \overline{L}\right). \tag{77}$$

This is true when $k = 0$, as both sides of (77) are $\lambda\left(L, \overline{L}\right)$. Let us suppose that the last vector field is $L_{k+1} = L$; the conjugated case is essentially the same. Then

$$\begin{aligned}
\langle \Upsilon_{k+3}, \eta \rangle &= \langle [\Upsilon_{k+2}, L], \eta \rangle \\
&= (\alpha_L - L)\langle \Upsilon_{k+2}, \eta \rangle - \lambda\left(L, \pi_{0,1}\Upsilon_{k+2}\right).
\end{aligned} \tag{78}$$

Since $T_p^{1,0}(M)$ is one-dimensional, we may write $\pi_{0,1}\Upsilon_{k+2} = \overline{f L}$ for some function f. Plugging this into (77) completes the induction step. The upshot is that both $c_L(p)$ and type$_p L$ equal the smallest value of $k+2$ such that the first term on the right in (78) does not vanish at p. Hence these invariants are equal in this case.

To relate them to the other numbers, we return to the extrinsic situation, and choose a defining function and coordinates such that p is the origin and

$$r(z, \bar{z}) = 2\operatorname{Re}(z_2) + f(z_1, \overline{z_1}) + \operatorname{Im}(z_2) g(z_1, \overline{z_1}, \operatorname{Im}(z_2)). \tag{79}$$

Here the functions satisfy

$$f(0) = df(0) = 0$$

$$g(0) = 0. \tag{80}$$

In these coordinates, $r \mid V$ is simply the function f. We may also suppose that the vector field is

$$L = \frac{\partial}{\partial z_1} - \frac{r_{z_1}}{r_{z_2}}\frac{\partial}{\partial z_2} \tag{81}$$

for it is easily seen that the two expressions involving L are unchanged if we multiply it by a nonvanishing function. Let T be the vector field

$$T = \frac{1}{r_{z_2}}\frac{\partial}{\partial z_2} - \frac{1}{\overline{r_{z_2}}}\frac{\partial}{\partial \overline{z_2}}. \tag{82}$$

With these choices we can even define the function α_L by

$$[T, L] = \alpha_L T \tag{83}$$

and the Levi form by

$$[L, \overline{L}] = \lambda\left(L, \overline{L}\right) T. \tag{84}$$

By using the explicit formula (64) for the Levi form, we see that

$$\lambda\left(L, \overline{L}\right) = \frac{r_{1\bar{1}}|r_2|^2 - 2\operatorname{Re}\left(r_{1\bar{2}}r_2 r_{\bar{1}}\right) + r_{2\bar{2}}|r_1|^2}{|r_2|^2}. \tag{85}$$

Keeping in mind also the form of L, it is easy to check that the first nonvanishing derivative of the Levi form at the origin occurs when all the derivatives are taken with respect to either z_1 or $\overline{z_1}$. This implies therefore the equivalence of 4 (and hence 3) with 5. We have seen already in the proof of Theorem 1 that the curve of maximal order of contact satisfies $z_2(t) = 0$ with the coordinates as chosen. Thus 5 is equivalent to 2, and in this case, there is no need to consider singular curves, so 2 is equivalent to 1. Let us remark why it is enough to consider nonsingular curves. Since the curve with maximal order of contact satisfies $z_2(t) = 0$, we may assume that it is given by $z_1(t) = t$, which is nonsingular. ∎

REMARK 1 For a given vector field on a CR manifold, it is not generally true that

$$c_p(L) = \text{type}_p L. \tag{86}$$

In case the CR manifold is pseudoconvex, the author [D6] has obtained partial results suggesting that this is always true. Bloom [Bl] has proved a related result for hypersurfaces in $\mathbb{C}^3$. The equality of these two numbers is stated and allegedly proved in [Si1] but the proof there is incorrect, as an identity that holds at only one point is differentiated. The argument there assumes that the terms arising from derivatives of the second term in (78) do not matter in the pseudoconvex case. Simple examples show that they are not error terms, but positivity considerations apparently prevent cancellation from occurring (see [D6]). Among the things proved there is that, in the pseudoconvex case, the two numbers are simultaneously four, even though these troublesome terms appear. ∎

4.3.2 Points of finite q-type

We have seen that a point of finite one-type is a point for which there is a bound on the maximum order of contact of parameterized (one-dimensional) complex analytic curves. Once one realizes that it is necessary to take into account the multiplicity of such a curve, it is clear what number deserves to be called the maximum order of contact. It is not so clear how to define the order of contact of higher dimensional varieties. From the point of view of Chapter 2 of this book, there is a natural way to do so. With this definition it is easy to extend the results of this chapter to this more general situation. In his work on subellipticity, Catlin has an alternative method for defining the order of contact for varieties of dimension greater than one.

The technique thus far has been to assign to a hypersurface the family of ideals $I(U, k, p)$ determined by the holomorphic decomposition of the kth-order Taylor polynomial of some defining function. To generalize to the case of higher dimensional varieties, we consider the effect of including more functions in these ideals, according to the ideas of Chapter 2. This process yields the following definitions.

DEFINITION 4   *The maximum order of contact of q-dimensional complex analytic varieties with a real hypersurface M at the point p is the number*

$$\Delta^q (M, p) = \inf_P \left\{ \Delta^1 (M \cap P, p) \right\}. \tag{87}$$

Here the infimum is taken over all choices of complex affine subspaces of dimension $n - q + 1$ passing through the point p.

Some simple remarks are appropriate. First, when q equals one, the definition is the same as before. When q equals $n - 1$, the definition reduces to the maximum order of tangency of the complex tangent hyperplane. Thus the $n - 1$ type can be computed by any of the methods that work in two complex variables. One restricts everything to a complex two-dimensional subspace and applies any of the methods from the previous section.

DEFINITION 5   *The q-multiplicity $B^q (M, p)$ is defined by*

$$B^q (M, p) = \inf_P \left\{ B (M \cap P), p \right\}, \tag{88}$$

where again the infimum is taken over all choices of complex affine subspaces of dimension $n - q + 1$ passing through the point p.

It is not difficult to extend the results about openness to this case. The reader can mimic the proof of Theorem 4 to prove the following result.

THEOREM 10
The set of points on a real hypersurface M for which

$$\Delta^q (M, p) < \infty \tag{89}$$

is an open subset of M. Also, $\Delta^q (M, p) < \infty \Leftrightarrow B^q (M, p) < \infty$.

Let us compute these numbers all these numbers for some algebraic real hypersurfaces. The reader should invent his own examples.

Example 5 Put

$$r (z, \bar{z}) = 2 \operatorname{Re} (z_5) + \left| z_1^2 - z_2^3 \right|^2 + \left| z_2^4 - z_3^5 \right|^2 + \left| z_4 \right|^2. \tag{90}$$

Assume the point in question is the origin. Observe first that the parameterized one-dimensional variety $t \to \left(t^{15}, t^{10}, t^8, 0, 0 \right)$ lies in M. This curve is irreducible and is defined once by the associated family of ideals. In the language of Chapter 2, the length of the module of functions on it equals unity. The existence of this curve shows that the origin is not a point of finite 1-type. The origin is a point of finite q-type for $q > 1$. Restricting the defining equation to the appropriate dimensional subspaces, and using the methods of Chapter 2, one easily computes the following values:

$$B^1(M,p) = \infty \qquad \Delta^1(M,p) = \infty$$

$$B^2(M,p) = 16 \qquad \Delta^2(M,p) = 8$$

$$B^3(M,p) = 4 \qquad \Delta^3(M,p) = 4$$

$$B^4(M,p) = 2 \qquad \Delta^4(M,p) = 2. \quad \square \tag{91}$$

4.3.3 Regular orders of contact

In the definition of order of contact of one-dimensional complex analytic varieties with a hypersurface, the obvious difficulty was the possibility that a variety could be singular. It is natural to ask what sort of theory one can develop by restricting consideration to complex manifolds. Analogs of the important results of this chapter no longer hold. In this section we first give the definition of regular order of contact. We then offer some examples where the conclusions of the theorems do not hold under such a finiteness hypothesis.

DEFINITION 6   The maximum order of contact of q-dimensional complex analytic manifolds with a real hypersurface M at a point p is the number

$$\Delta_{\text{reg}}^q(M,p) = \sup\left(v\left(z^*r\right)\right) \tag{92}$$

where r is a local defining function for M and

$$z : \left(\mathbb{C}^q, 0\right) \rightarrow \left(\mathbb{C}^n, p\right) \tag{93}$$

is the germ of a holomorphic mapping for which $\operatorname{rank}(dz(0)) = q$.

A simple verification shows that the order of contact is independent of the choice of the defining function. There are other possible definitions, of course; we choose this one because of its similarity with the definition of order of contact of one-dimensional varieties. This number also goes by the name "regular order of contact." The following example convinced the author that the concept fails to describe the geometry adequately.

Example 6

Put

$$r(z, \bar{z}) = 2 \operatorname{Re}(z_3) + \left| z_1^2 - z_2^3 \right|^2. \tag{94}$$

Then the regular order of contact at the origin is easily seen to be 6. Yet there are points close by for which the regular order is infinite; this holds at every smooth point of the variety

$$\mathbf{V}\left(z_3, z_1^2 - z_2^3\right). \tag{95}$$

Thus the function

$$p \rightarrow \Delta_{\text{reg}}^1(M, p) \tag{96}$$

is not locally bounded, and the set of points of finite regular one-type is not an open subset. ⬜

REMARK 2 In his work on subellipticity, Catlin assigns a multi-type to a point. The multi-type is an n-tuple of rational numbers. For this example, the multi-type at the origin is the triple $(1, 4, 6)$, while at nearby points it is the triple $(1, 2, \infty)$. Catlin proves that the multi-type is upper semicontinuous in the lexicographic ordering. The naive concept of regular order of contact cannot detect that things may be more singular nearby. Catlin's multi-type does, in this lexicographic sense. The approach in this book makes it possible to see exactly what is happening at all nearby points. ∎

REMARK 3 It is not difficult to prove that the regular order of contact agrees with the order of contact when the regular order is at most 4. We sketch this in Proposition 3. It is even easier to prove that the concepts agree when $q = n - 1$, that is,

$$\Delta^{n-1}(M, p) = \Delta_{\text{reg}}^{n-1}(M, p). \tag{97}$$

To do so, one notes that the intersection taken in Definition 4 reduces the problem to finite type in two dimensions. In two dimensions the result is part of Theorem 9. When $q = n - 1$, the multiplicity also has the value (97), for essentially the same reason. It is this phenomenon that explains all the successes for domains in two dimensions, or for results concerning extensions of CR functions ($n - 1$ type arises here). In both of these cases only regular order of contact is needed. There are other general classes of domains with simple geometry for which the regular order of contact equals the order of contact. McNeal [Mc1,Mc2] proved this statement for convex domains and used it in a study of the boundary behavior of the Bergman kernel function. ∎

PROPOSITION 3

Let M be a real hypersurface in $\mathbb{C}^n$. Suppose that $\Delta^1_{\text{reg}}(M, p) \leq 4$. Then $\Delta^1(M, p) = \Delta^1_{\text{reg}}(M, p)$.

SKETCH OF PROOF Without loss of generality we may assume that p is the origin. Let r be a local defining equation. Suppose that we are given a curve z with $v(z) = m$. We can then write

$$z(t) = (t^m w(t)) \tag{98}$$

where $w(0) \neq 0$. Suppose that $v(z^*r) > 4v(z) = 4m$, that is, $\Delta^1(M, p) > 4$. We differentiate the equation

$$r(t^m w(t)) = r(z(t)) = 0\left(|t|^{4m+1}\right) \tag{99}$$

with respect to both t and $\bar{t}$ up to a total of at most $4m$ times, and then we evaluate at the origin. We obtain a large number of homogeneous multilinear equations in the derivatives of z. Since all the derivatives of z up to order $m - 1$ vanish, many of these equations are trivial. A way to take advantage of this is to notice at the start that there is no loss in generality in assuming that r is a polynomial of degree 4, when differentiating (99). After listing all the nontrivial equations, one sees that among them are ten equations that involve the mth derivative of z but no higher derivatives. These ten equations involve the ten possibilities obtained where the coefficients are $(D^a \overline{D}^b r)_0$ for $1 \leq a + b \leq 4$. Consider on the other hand the complex line defined by $t \rightarrow tw(0) = \zeta(t)$. These ten equations say precisely that

$$D^a \overline{D}^b \left(\zeta^* r\right)_0 = 0 \qquad 1 \leq a + b \leq 4. \tag{100}$$

Formula (100) shows, however, that the nonsingular curve ζ has order of contact greater than four. Since it is nonsingular, we obtain the following conclusion. Whenever $\Delta^1(M, p) > 4$, necessarily $\Delta^1_{\text{reg}}(M, p) > 4$ also. The hypothesis of the proposition therefore implies that $\Delta^1(M, p) \leq 4$. In a similar fashion we see that a similar result holds for orders 2 and 3. Putting this information together yields the desired conclusion. ∎

4.4 Conclusions

We have assigned many numbers to a point of a real hypersurface. All of these numbers can be computed by algebraic methods. More important than the numbers themselves are the families of ideals $I(U, k, p)$. From these ideals we can estimate all of the numbers, and in the pseudoconvex case, from these ideals we can compute them all exactly. It is valuable to remark that a point on any hypersurface is of type two precisely when the hypersurface is strongly pseudoconvex. In that case all the ideals are the maximal ideal. If the hypersurface has (even a nondegenerate) Levi form with eigenvalues of both signs, then the linear terms in the g's cancel the linear terms in the f's and we obtain a point of higher type. Points where the Levi form degenerates must be also of higher type.

It is useful to put together all of the inequalities relating the various numbers. We have proved the following results.

THEOREM 11

Let M be a real hypersurface containing the point p. For the one-type we have the estimates

$$\Delta^1_{\text{reg}}(M,p) \leq \Delta^1(M,p) \leq B^1(M,p) \leq 2\left(\Delta^1(M,p) - 1\right)^{n-1}. \tag{101}$$

For $1 \leq q \leq n-1$ we have

$$\Delta^q_{\text{reg}}(M,p) \leq \Delta^q(M,p) \leq B^q(M,p) \leq 2\left(\Delta^q(M,p) - 1\right)^{n-q}. \tag{102}$$

For the special case where $q = n-1$ we have the equalities

$$\Delta^{n-1}_{\text{reg}}(M,p) = \Delta^{n-1}(M,p) = B^{n-1}(M,p). \tag{103}$$

If we append to our hypotheses the statement that the Levi form has s eigenvalues of the same sign, then we can improve the exponent in the last column of (102) from $n-q$ to $n-q-s$ as long as $q \leq n-s$. If M is pseudoconvex, then the estimates can be improved as follows:

$$\Delta^1_{\text{reg}}(M,p) \leq \Delta^1(M,p) \leq B^1(M,p) \leq 2\left(\frac{\Delta^1(M,p)}{2}\right)^{n-1}$$

$$\Delta^q_{\text{reg}}(M,p) \leq \Delta^q(M,p) \leq B^q(M,p) \leq 2\left(\frac{\Delta^q(M,p)}{2}\right)^{n-q}$$

$$\Delta^{n-1}_{\text{reg}}(M,p) = \Delta^{n-1}(M,p) = B^{n-1}(M,p) \tag{104}$$

These estimates are sharp. We can improve them as above for each positive eigenvalue of the Levi form. Let $X^q(M,p)$ denote any of the three invariants. Essentially by definition we have

$$X^{n-1}(M,p) \leq X^{n-2}(M,p) \leq \cdots \leq X^1(M,p). \tag{105}$$

Each of the numbers $B^q(M,p)$ is upper semicontinuous as a function of the point or as a function of the hypersurface. Hence the far right side of (104) gives a local bound for the numbers on the left in an open neighborhood of p.

Thus

$$X^q(M, z) \leq 2 \left(\frac{\Delta^1(M, p)}{2} \right)^{n-q} \tag{106}$$

for all z near p.

In Theorem 2.5 we summarized the properties of numerical invariants of holomorphic ideals. In Chapters 3 and 4 we showed how to reduce the geometry of real hypersurfaces to properties of these holomorphic ideals. In Theorem 11 we have accomplished our aim in finding an appropriate generalization of Theorem 2.5 to real hypersurfaces in $\mathbb{C}^n$.

5

Proper Holomorphic Mappings Between Balls

5.1 Rational proper mappings

5.1.1 Introduction

In this chapter we continue our study of the geometry of real hypersurfaces by focusing on the unit sphere and on holomorphic mappings that send spheres to spheres. We will see rather different applications of some of the same ideas. In particular we will discover in Corollary 7 a remarkable fact. There are CR manifolds, namely quotients of spheres, from which there exist continuous nonconstant CR mappings to spheres, but from which there do not exist smooth nonconstant CR mappings to any sphere. This corollary follows from our study of invariant proper holomorphic mappings between balls. Another recurring theme in this chapter is the m-fold tensor product, a mapping that generalizes the function $z \to z^m$ in one dimension.

Recall that a holomorphic mapping $f : \Omega_1 \to \Omega_2$ is proper if $f^{-1}(K)$ is a compact subset of Ω_1 whenever K is a compact subset of Ω_2. In general there are no proper holomorphic mappings between a given pair of domains. A necessary condition is that the dimension of the image domain Ω_2 must be at least as large as the dimension of Ω_1. The easiest way to see this is to first observe that the inverse image of a point must be a compact complex analytic subvariety, and hence a finite set. If a proper mapping decreased dimensions, however, then the inverse image of a point would have to be a complex analytic subvariety of positive dimension.

The first conceivable case of interest is therefore the equidimensional one. For a given pair of domains in the same dimension it remains generally impossible to find a proper mapping between them. On the other hand, given one domain, it often can be embedded properly into a ball or polydisc of sufficiently high dimension. We discuss this briefly in Section 3.3.

The primary purpose of this chapter is to discuss proper mappings between balls; these are of interest in part because of their abundance and in part because the symmetry of the ball allows for interesting mathematics. Also one can get quite far with simple ideas. This is evident in the one-dimensional case, where we have the simple classical theorem that every proper holomorphic mapping of the unit disc is a finite Blaschke product. In higher dimensions, but where the domain and target balls are of the same dimension (the equidimensional case), proper mappings are necessarily automorphisms, and thus their components are again rational functions. We will use the term codimension here to mean the difference of the target and domain dimensions. When one considers proper mappings in the positive codimension case, additional considerations arise. Assuming enough boundary smoothness, such mappings are necessarily rational. In every positive codimension, however, there are proper mappings, continuous on the closed ball, whose components are not rational functions.

In most of this chapter we restrict consideration to rational proper holomorphic mappings between balls. Such mappings give a higher dimensional generalization of Blaschke products. The generalization of ordinary multiplication turns out to be the tensor product on a subspace. We also consider invariance properties of mappings under fixed-point–free finite unitary groups. In this regard we solve completely the problem for which such groups do there exist invariant proper rational mappings.

5.1.2   Blaschke products

The purpose of this section is to find all proper self-mappings of the unit disc. The result is quite easy.

THEOREM 1

Every proper holomorphic mapping f from the unit disc to itself is a finite Blaschke product. That is, there are finitely many points $\{a_j\}$ in the disc, positive integer multiplicities m_j, and a number $e^{i\theta}$ such that

$$f(z) = e^{i\theta} \prod_{j=1}^{m} \left(\frac{a_j - z}{1 - \overline{a}_j z} \right)^{m_j}. \tag{1}$$

PROOF   First consider the set $f^{-1}(0)$. It is compact because the mapping is proper, and it is a (zero-dimensional) complex analytic variety. Therefore it is a discrete set of points $\{a_j\}$ with corresponding multiplicities m_j. Denote by $b(z)$ the product in (1). Each factor in the product is an automorphism of the disc and maps the circle to itself. Thus the product maps the disc to itself and the circle to itself, and is therefore a proper mapping. To show that $f(z) = e^{i\theta} b(z)$, it is enough to show that $|f(z)| = |b(z)|$. (Note the analogy with Proposition 3.3.) Each of the functions f/b and b/f has only removable singularities, and hence is holomorphic. Using an $\epsilon - \delta$ version of Lemma 1.1, we find that, given ϵ, there is a δ so that

$$1 - \epsilon \leq \left| \frac{f(z)}{b(z)} \right| \leq \frac{1}{1 - \epsilon} \tag{2}$$

whenever $|z| > 1 - \delta$. By the maximum principle applied to both f/b and b/f, we see that inequality (2) persists on an open disc about 0. Since ϵ is arbitrary we can conclude that f/b has constant modulus, and the result follows. ∎

Thus the proper mappings of the disc are easy to classify. They are all rational functions that extend holomorphically past the disc. They can have any finite number of zeroes with any finite multiplicities at these zeroes. To get an analogous result in several variables, we cannot stay in the equidimensional situation. This is because of a result essentially known to Poincaré (see [Fo3] for the history) that was proved by Alexander [A].

5.1.3 The equidimensional case

Let us denote by B_n the unit ball in $\mathbb{C}^n$. In the middle 1970s Alexander [A] proved that proper mappings from B_n to itself are necessarily automorphisms when n is at least two. His work inspired much progress. It is now known that proper holomorphic self-mappings of strongly pseudoconvex domains (or of bounded domains with real analytic boundaries) must be automorphisms in two or more dimensions. See the articles [Bed,Pi2]. Returning to proper self-mappings of the ball, one notes that it is impossible to prescribe multiplicities as in the one-dimensional case. Later we will see that one can do so by allowing the target ball to be of sufficiently large dimension.

The automorphism group $\mathrm{Aut}(B_n)$ of the ball has the following description. [Ru1]. Every f in $\mathrm{Aut}(B_n)$ is of the form

$$f = U \xi_a \tag{3}$$

where U is unitary, and ξ_a is the linear fractional transformation

$$\xi_a(z) = \frac{a - L_a(z)}{1 - \langle z, a \rangle} . \tag{4}$$

Here L_a is a linear transformation defined by

$$L_a(z) = sz + \frac{\langle z, a \rangle}{s + 1} a \tag{5}$$

and s is the positive number satisfying $s^2 = 1 - ||a||^2$. One could also describe $\mathrm{Aut}(B_n)$ as the real Lie group $\mathbf{SU}(n, 1)$.

There are several proofs of the theorem of Poincaré and Alexander and many generalizations, especially by Pinchuk [Pi1]. We present now a proof that is due essentially to Bell [Be1]. See also his survey article [Be3] for an elementary exposition of this proof. We begin with a lemma that holds in more generality than for proper mappings between balls. In this lemma we do not assume that the domains lie in the same dimensional spaces.

LEMMA 1
Suppose that $f : \Omega_1 \to \Omega_2$ is a proper holomorphic mapping between strongly pseudoconvex domains that extends to be twice continuously differentiable at the boundary. Then $df(p)$ has maximal rank at each point p in $b\Omega_1$.

PROOF   Note first that if $\Omega_1 \subset \mathbb{C}^n$, $\Omega_2 \subset \mathbb{C}^n$, and a proper holomorphic mapping between them exists, then necessarily $N \geq n$. We may choose coordinates so that $p = 0$, $f(p) = 0$. After a linear change of coordinates we suppose that we are given defining equations of the form

$$r(z) = 2\,\mathrm{Re}\,(z_n) + \phi(z)$$

$$\rho(w) = 2\,\mathrm{Re}\,(w_N) + \psi(w) \tag{6}$$

where the functions ϕ, ψ vanish to order two at the origin. We may suppose further by the strong pseudoconvexity that the defining function ρ is strongly plurisubharmonic. The function $z \to \rho(f(z))$ is then also plurisubharmonic, negative on Ω_1, and achieves its maximum value of zero at the origin, since f is proper. By the Hopf lemma (see [Ru1] for example), its outer normal derivative must be strictly positive. The form of the defining equations show that this outer normal derivative is precisely $(\partial f_N/\partial z_n)(0)$. In particular, the function $z \to \rho(f(z))$ is then a local defining function for Ω_1. As such, its Levi form is positive definite on the space $T^{1,0}(b\Omega_1)$. On the other hand, this defining function is also a multiple of r. We differentiate and evaluate at the origin. Since $(\partial r/\partial z_j)(0) = 0$ for $j < n$, we see that

$$\rho(f) = gr \Rightarrow \frac{\partial f_N}{\partial z_j}(0) = 0 \qquad j < n. \tag{7}$$

Suppose now that $df(0)v = 0$ for some $(1,0)$ vector v in $\mathbb{C}^n$. Then $v_n = 0$ by (7), and hence v is tangent to $b\Omega_1$. Computation by the chain rule of the Levi form $\partial\bar{\partial}(\rho(f))(v, \bar{v})$ on v at the origin shows that

$$\partial\bar{\partial}(\rho(f))(v, \bar{v}) = \partial\bar{\partial}\rho\left(\partial f(0)(v), \overline{\partial f(0)(v)}\right) = 0. \tag{8}$$

Thus the vector v lies in the kernel of a positive definite form, and hence it is zero. Thus the kernel of $df(0)$ is trivial, and we obtain the desired conclusion.   ∎

THEOREM 2
A proper holomorphic self-mapping of B_n is necessarily an automorphism when $n \geq 2$.

PROOF The first step is to show that the proper mapping must extend smoothly to the boundary. Once this is accomplished, it follows from Lemma 1 that the Jacobian determinant $\det(df(z))$ cannot vanish for $z \in \mathrm{b}B_n$. This implies that $\det(df(z))$ cannot vanish anywhere on B_n, by the following argument. Since

$$z \to \det(df(z)) \tag{9}$$

is holomorphic on B_n, its zero set is a complex analytic subvariety of the ball of dimension $n - 1 > 0$. Since there are no compact positive dimensional complex analytic subvarieties of the ball [Ru1], we see that this variety must be empty. (If not, then it would extend to the boundary, but we know $\det(df(z))$ does not vanish there.) Once we know that $\det(df(z))$ never vanishes, f is an unbranched covering mapping of a simply connected domain, and therefore injective. To finish the proof of the theorem, it remains to verify that the mapping extends smoothly to the boundary. The smooth extendability to the boundary is a special case of Theorem 7.2, but can be proved directly for the ball by way of the explicit formula for the Bergman kernel function (see [Be1]). ∎

There is some value in considering the proof of Lemma 1 in the special case of the ball. The conditions of properness and extendability guarantee that

$$\Psi = \rho(f) = \|f\|^2 - 1 = 0 \quad \text{on} \quad \|z\|^2 = 1. \tag{10}$$

As above, it follows from the Hopf lemma that $d\Psi = d(\|f\|^2) \neq 0$ on the sphere and hence $\|f\|^2 - 1$ is a defining function for the ball. Since the ball is strongly pseudoconvex, its Levi form is positive definite on $T^{1,0}\mathrm{b}B_n$. Computation of the necessarily nonvanishing determinant of the Levi form shows that it has $|\det(df(z))|^2$ as a factor.

There is a pleasing alternative manner of phrasing Lemma 1. A proper mapping between strongly pseudoconvex domains that is smooth up to the boundary defines, when restricted to the boundary, a CR immersion of one strongly pseudoconvex CR manifold into another.

REMARK 1 It is possible in the equidimensional case to avoid the use of the Hopf lemma to obtain one nonvanishing derivative. There is a formal power series argument of Baouendi and Rothschild [BR2] that holds when the target domain is of finite $(n-1)$-type and when the domains are in the same dimension. In the notation of Lemma 1, this argument shows that $(\partial f_N / \partial z_n)(0) \neq 0$. Without strong pseudoconvexity it is impossible to conclude that any more derivatives must not vanish. ∎

5.1.4 Polynomial proper mappings between balls

When the author first heard the statement of Theorem 2, he wondered what happened to functions like z^m. In order to obtain an analogue of this function, one must consider the tensor product. Since this operation increases the target dimension, it turns out that one must study proper mappings from a given ball to all possible larger dimensional balls to see the appropriate analogues of finite Blaschke products. This material is the subject of this section.

In particular, we classify all proper polynomial mappings between balls. The key idea is that the operation of tensor product on a subspace generalizes the notion of multiplication. We also must allow the undoing of a tensor product. Suppose that g, h are holomorphic mappings with the same domain Ω, but taking values in perhaps different finite-dimensional complex vector spaces $\mathbb{C}^n$ and $\mathbb{C}^N$. We then define the tensor product of the mappings by

$$g \otimes h = (g_1 h_1, g_2 h_1, \ldots, g_n h_1, g_1 h_2, \ldots, g_n h_N). \tag{11}$$

Thus the tensor product of two mappings is the mapping whose components are all possible products of the components of the two mappings. In the present context, proper mappings between balls that agree up to unitary transformations will be considered the same. Thus it does not really matter what order we assign to the list of components. What we do need is the possibility of applying this operation on subspaces.

Suppose that we have written a complex vector space as an orthogonal sum

$$\mathbb{C}^N = A \oplus A^\perp. \tag{12}$$

If f is a holomorphic mapping that takes values in $\mathbb{C}^N$, then it has then an induced decomposition

$$f = f_A \oplus f_{A^\perp}. \tag{13}$$

Suppose also that g is any holomorphic mapping with the same domain as f, and with range contained in any finite-dimensional complex vector space. Given an orthogonal summand A, we can form the mapping

$$E(A, g)(f) = (f_A \otimes g) \oplus f_{A^\perp}. \tag{14}$$

We also have the undoing operation, defined on mappings in the range of (14), by

$$E^{-1}(A, g)((f_A \otimes g) \oplus f_{A^\perp}) = f. \tag{15}$$

LEMMA 2
Suppose that

$$f : \Omega \subset\subset \mathbb{C}^n \to B_N$$

$$g : \Omega \subset\subset \mathbb{C}^n \to B_K \tag{16}$$

are proper holomorphic mappings. Suppose that the range of f has the orthog-
onal decomposition $\mathbb{C}^N = A \oplus A^\perp$. *Then the mapping*

$$E(A, g)(f) = (f_A \otimes g) \oplus f_{A^\perp} \qquad (17)$$

defines a proper holomorphic mapping from Ω *to the appropriate dimensional ball.*

PROOF Compute first that

$$\begin{aligned} \|E(A, g)(f)\|^2 &= \|f_A \otimes g\|^2 + \|f_{A^\perp}\|^2 \\ &= \|f_A\|^2 \|g\|^2 + \|f_{A^\perp}\|^2. \end{aligned} \qquad (18)$$

Now let the point z tend to $b\Omega$. Since g is proper, $\|g\|^2$ tends to unity. Thus $\|E(A, g)(f)\|^2$ has the same limit as $\|f_A\|^2 + \|f_{A^\perp}\|^2 = \|f\|^2$ which is also unity because f is proper. Thus the result follows from Lemma 1.1. ∎

The tensor product serves as the appropriate generalization of ordinary multiplication in one dimension. It is interesting that one must also allow the
"undoing" of a tensor product in higher dimensions; this analogue of division
is superfluous in one variable. In higher dimensions it arises because one may
wish to "undo" on a different subspace from the one on which one "did."

Let us give some examples. The simplest case where one must "undo" is the
following:

Example 1
Put

$$f(z, w) = (z, w)$$

$$A = \operatorname{span}\{(1, 0)\}$$

$$g(z, w) = (z, w). \qquad (19)$$

Performing the tensor product on this subspace yields the mapping

$$(E(A, g) f)(z, w) = h(z, w) = \left(z^2, zw, w\right). \qquad (20)$$

Now decompose the three-dimensional range by choosing the subspace generated by $(1, 0, 0)$ as A and again tensor with the mapping g, yielding the mapping
$h_2(z, w) = \left(z^3, z^2 w, zw, w\right)$. Several more applications eventually yield

$$H_3(z, w) = \left(z^3, z^2 w, z^2 w, z^2 w, zw^2, zw^2, zw^2, w^3\right). \qquad (21)$$

Now consider a unitary change of coordinates in this eight-dimensional space and obtain

$$H(z, w) = \left(z^3, \sqrt{3}z^2 w, \sqrt{3}zw^2, w^3, 0, 0, 0, 0 \right). \tag{22}$$

After identifying (22) with its projection onto the first four coordinates, we obtain a proper holomorphic mapping to the four-dimensional ball. The second and third components are in the range of the tensor product operation. After undoing, we obtain

$$G(z, w) = \left(z^3, \sqrt{3}zw, w^3 \right). \tag{23}$$

The point of this example is that the "undoing" is required. It is easy to prove that the mapping (23) cannot be constructed from tensor products alone.   ☐

The homogeneous mapping (21) can be reached more quickly by tensoring on the full space three times. This technique is important in the general theory; see Example 2. We now let $z = (z_1, \ldots, z_n)$ denote the domain variable. Often it is important to assume that the target dimension is minimal; this applies to (22). Henceforth we allow the slight abuse of notation that occurs by identifying the mapping F defined by $F(z) = (f(z), 0)$ with f, even though these mappings have different ranges.

Example 2

(Homogeneous proper mappings between balls). Suppose that we perform the tensor product of the identity mapping with itself, always on the full space, a total of m times. We obtain a mapping whose components are all monomials of degree m, with each component being listed the appropriate number of times. After performing a unitary change of coordinates that collects the same monomials together, and ignoring zero components, one obtains the mapping

$$H_m(z) = z \otimes z \otimes \cdots \otimes z \qquad m\text{-times}. \tag{24}$$

It defines a homogeneous monomial mapping of degree m, namely,

$$H_m(z) = (\ldots, c_a z^a, \ldots)$$
$$|c_a|^2 = \binom{m}{a} = \frac{m!}{a_1! \ldots a_n!} \tag{25}$$

and turns out to be particularly important in the classification of proper mappings between balls.   ☐

The importance of the mapping $z \to H_m(z)$ derives in part from the following simple result, noticed independently by Rudin [Ru2] and the author [D8]. This result implies that H_m is essentially the only homogeneous proper mapping between balls. This result is important in our classification, so we give several simple proofs.

THEOREM 3
Suppose that

$$f : B_n \to B_N \tag{26}$$

is a proper mapping between balls, the components of f are linearly independent, and each component is a homogeneous polynomial of degree m. Then, after composition in the range with a unitary transformation, f is the mapping H_m.

PROOF   As a proper mapping from ball to ball that extends past the boundary, f must map the sphere to itself. This yields the equations

$$\|f(z)\|^2 = 1 \qquad \text{on} \quad \|z\|^2 = 1 \Rightarrow$$
$$\|f(z)\|^2 = \|z\|^{2m} \quad \text{on} \quad \|z\|^2 = 1. \tag{27}$$

The second equality is one of homogeneous polynomials, so it holds in all of $\mathbb{C}^n$. Using the multinomial expansion, this equality becomes

$$\|z\|^{2m} = \sum_{|a|=m} \binom{m}{a} |z|^{2a} = \|H_m(z)\|^2. \tag{28}$$

Note the use of multi-index notation in the middle term. We are now in the situation of Proposition 3.3, where we have two holomorphic functions with the same squared norm. According to that proposition, they differ by a unitary matrix of constants. Thus, for some U,

$$f(z) = U H_m(z). \tag{29}$$

This completes the first simple proof.   ∎

It is convenient to sketch another simple proof. First, it is not hard to show that, after composing with a linear transformation, we may assume that the components of the mapping are linearly independent monomials $c_a z^a$. We replace the variable z by ζ/s and use the homogeneity to obtain

$$\sum_{|a|=m} |c_a|^2 |\zeta|^{2a} = |s|^{2m} = \|\zeta\|^{2m} = \sum_{|a|=m} \binom{m}{a} |\zeta|^{2a}. \tag{30}$$

Again the result follows upon equating coefficients. (Note again the use of multi-index notation.) A polarized version of this proof appears in the section on varieties associated with proper mappings, where we show that, for this particular mapping, the associated variety equals the graph of the mapping.

By a process of "orthogonal homogenization," one can reduce the classification of proper polynomial mappings between balls to Theorem 3. Suppose now
that we have an arbitrary polynomial proper mapping of degree m. Let us write

$$f(z) = \sum_{|a| \leq m} v_a z^a = \sum_{k=0}^{m} f_k(z) \tag{31}$$

where each coefficient v_a is a point in $\mathbb{C}^n$ and each f_k is a homogeneous
(vector-valued) polynomial of degree k. We can use either form to derive
useful information from the relation that

$$\|f(z)\|^2 = 1 \quad \text{on} \quad \|z\|^2 = 1. \tag{32}$$

We derive these identities in the more general rational case, as we will use
them there also, and the technique is the same. Let us suppose that $f = p/q$ is
a rational function that satisfies (32). Write

$$p(z) = \sum_{|\alpha|=0}^{s} c_\alpha z^\alpha = \sum_{k=0}^{s} p_k$$

$$q(z) = \sum_{|\alpha|=0}^{s} d_\alpha z^\alpha = \sum_{k=0}^{s} q_k \tag{33}$$

and substitute (33) into (32). Clear denominators. Replace z by $re^{i\theta}$ (this is
multi-index notation) in the result. Equate Fourier coefficients. We obtain the
identities

$$0 = \sum_{|\beta|=0}^{s} \left(\langle c_{\beta+\gamma}, c_\beta \rangle - d_{\beta+\gamma}\overline{d_\beta} \right) r^{2\beta} \quad \forall \gamma \tag{34}$$

on the set $\|r\|^2 = 1$. We can do the same for the homogeneous parts. Replace
z by $ze^{i\theta}$ where now θ denotes just one variable. Again by equating Fourier
coefficients we obtain

$$0 = \sum_{k=0}^{s} \left(\langle p_{k+j}, p_k \rangle - q_{k+j}\overline{q_k} \right) \quad 0 \leq j \leq s \tag{35}$$

on the sphere. These identities can then be homogenized so as to hold everywhere. Doing this to (35) yields

$$\sum_{k=0}^{s} \left(\langle p_{k+j}, p_k \rangle - q_{k+j}\overline{q_k} \right) \|z\|^{2s-2k} = 0 \quad \forall j. \tag{36}$$

In the polynomial case, we may assume without loss of generality that $q = 1$.

Putting $f = \sum f_k$, we obtain from (35) the following. On the sphere

$$\sum_{k=0}^{m-j} \langle f_{k+j}, f_k \rangle = 0 \qquad j = 1, \ldots, m$$

$$\sum_{k=0}^{m} \|f_k\|^2 = 1. \tag{37}$$

From (37) we obtain identities on all of $\mathbb{C}^n$:

$$\sum_{k=0}^{m-j} \langle f_{k+j}, f_k \rangle \|z\|^{2m-2k} = 0 \qquad j = 1, \ldots, m$$

$$\sum_{k=0}^{m} \|f_k\|^2 \|z\|^{2m-2k} = \|z\|^{2m}$$

$$\sum_{k=0}^{m} \|f_k \otimes H_{m-k}\|^2 = \|H_m\|^2. \tag{38}$$

The third statement is alternate notation for the second and suggests that a result such as Theorem 4 might hold. This result shows that every polynomial proper mapping between balls can be constructed via the same algorithm; there are choices of subspaces involved that determine the mapping. This enables us to completely classify the polynomial proper mappings between balls in all dimensions.

THEOREM 4

Every proper holomorphic polynomial mapping f of degree m between balls is constructed in the following manner. Begin with the (essentially unique) homogeneous proper mapping H_m. Then there exists a finite list of subspaces such that

$$f = \prod_{j=0}^{m} E^{-1}(A_j, z) H_m. \tag{39}$$

REMARK 2 The notation in (39) does not preclude the possibility that some of the subspaces A_j may be trivial. ∎

PROOF Let us write $f = \sum_{k=0}^{m} f_k$ for the expansion of f into homogeneous polynomials. According to (37), we must have $0 = \langle f_0, f_m \rangle$. Therefore we can choose an orthogonal decomposition

$$\mathbb{C}^N = A \oplus A^{\perp} \tag{40}$$

such that $f_0 \in A$, $f_m \in A^{\perp}$. Note that this means that $f_m(z) \in A^{\perp} \; \forall z$. For any such decomposition, we can form the new mapping

$$E(A, z) f = (f_A \otimes z) \oplus f_{A^\perp}. \tag{41}$$

According to Lemma 2, $E(A, z) f$ defines a proper mapping with domain the same ball but with range a perhaps larger dimensional ball. It is evidently also still a polynomial. The important things are that (41) is still of the same degree, and that it has no constant term. Suppose inductively that we have a mapping $g = \sum_{k=s}^m g_k$ where s is larger than zero but less than m. The orthogonality relations then imply that

$$\langle g_s, g_m \rangle = 0. \tag{42}$$

Again we can choose an orthogonal decomposition of the range and apply the tensor product. Doing so yields a new mapping

$$E(A, z) g = (g_A \otimes z) \oplus g_{A^\perp}. \tag{43}$$

The important thing now is that (43) remains of degree m, but has no term of degree s. Thus we have obtained a proper mapping that satisfies the same hypotheses except that it vanishes to one order higher. This process continues until the mapping is homogeneous. On the other hand, after subjecting a homogeneous proper mapping to a linear transformation, it is essentially uniquely determined. This shows that

$$H_m(z) = \prod_{j=0}^m E\left(A_j, z\right) f. \tag{44}$$

Putting all the factors (in opposite order) of E^{-1} on the other side of (44) yields the desired conclusion. ∎

Since the homogeneous mapping H_m is itself a tensor product, we can state the result in Theorem 4 as a composition product factorization. Using the notation (14), where each $A^\perp$ is trivial, we can write (24) as

$$H_m(z) = z \otimes \cdots \otimes z = \prod_{j=1}^m E\left(V_j, z\right)(1) \tag{45}$$

for appropriate subspaces. This yields the next corollary.

COROLLARY 2
Every proper polynomial mapping between balls admits the composition product factorization

$$f(z) = \prod_{j=0}^m E^{-1}\left(A_j, z\right) L \prod_{j=1}^m E\left(V_j, z\right)(1). \tag{46}$$

Here 1 denotes the constant mapping to the point 1 on the unit circle, each V_j is the range of the mapping on which the tensor product is applied, L is linear, and each A_j is a subspace of the range of the mapping on which the tensor product is "undone."

We emphasize that each intermediate step defines a proper mapping to some ball. Let us denote by Ξ_n the collection of proper polynomial mappings with domain equal to B_n and range equal to some B_k. The three operations involved (composition with certain linear transformations, tensor products on subspaces of the range, and the undoing of the tensor product) map Ξ_n to itself. Under composition the collection of these operations on Ξ_n form a monoid. It is possible in explicit examples to write down the factorization given by this proof of Theorem 4 and Corollary 2. It may also be possible to write down a simpler one. Thus "unique factorization" fails. To show this, we factorize the simplest nontrivial case in two ways. Consider $f : B_2 \to B_3$ defined by

$$f(z_1, z_2) = \left(z_1^2, z_1 z_2, z_2\right). \tag{47}$$

To factorize it according to the proof, we write

$$\mathbb{C}^3 = \mathbb{C}^2 \oplus \mathbb{C} = \{s, t, 0\} \oplus \{0, 0, u\}. \tag{48}$$

Then the mapping is an orthogonal summand, and applying one tensor product we obtain that

$$f = \left(z_1^2, z_1 z_2, 0\right) \oplus (0, 0, z_2) \Rightarrow$$
$$E(A, z) f = \left(z_1^2, z_1 z_2, 0\right) \oplus ((z_1, z_2) \otimes (0, 0, z_2))$$
$$= \left(z_1^2, z_1 z_2, 0, 0\right) \oplus \left(0, 0, z_1 z_2, z_2^2\right)$$
$$= \left(z_1^2, z_1 z_2, z_1 z_2, z_2^2\right) \simeq H_2(z_1, z_2). \tag{49}$$

Note that $H_2(z)$ factors into two tensor products. Thus the proof of the theorem gives a factorization of the form

$$f = E(A_3, z)^{-1} E(A_2, z) E(A_1, z). \tag{50}$$

One can write

$$f = E(A, z)(z)$$
$$= ((z_1, 0) \otimes (z_1, z_2)) \oplus (0, 0, z_2) \tag{51}$$

for a simpler factorization.

REMARK 3 The same ideas apply in the rational case. By allowing also tensor products of automorphisms that move the origin, or composition with such automorphisms, one can factorize a rational proper mapping between balls in this manner. The proof is considerably more difficult to write down, as the analogue of the homogeneous mapping is a rational proper mapping of the form

$$\frac{p_{m-d} + p_{m-d+1} + \cdots + p_m}{1 + q_1 + \cdots + q_d} \tag{52}$$

where the individual terms are homogeneous polynomials. The reduction to this case is virtually identical to the reduction of a polynomial mapping to the homogeneous case. It is much harder to factorize and classify mappings such as (52) than to do the same for the homogeneous polynomial mapping (24) of Theorem 3. ∎

5.1.5 Multidimensional families

We begin by discussing the notion of spherical equivalence. Suppose that f, g : $B_n \rightarrow B_N$ are proper holomorphic mappings. They are called "spherically equivalent" if there are automorphisms $\varphi \in \text{Aut}(B_n)$, $\chi \in \text{Aut}(B_N)$ such that $g\varphi = \chi f$. They are called "unitarily equivalent" if these automorphisms can be chosen to be linear, hence unitary. In the same sense that one diagonalizes a linear transformation by appropriate choices of bases, one can try to put a proper mapping into a simpler form to which it is spherically equivalent. The simplest example of this is the following elementary observation.

REMARK 4   Every Blaschke product with exactly two factors is spherically equivalent to the function $z \rightarrow z^2$. The proof is left to the reader. ∎

Faran [Fa2,Fa1] has proved the following results on spherical equivalence.

THEOREM 5
Suppose that $f : B_n \rightarrow B_N$ is proper, smooth on the closed ball, and $N < 2n - 1$. Then f is spherically equivalent to the linear embedding given by $z \rightarrow (z, 0)$.

THEOREM 6
Suppose that $f : B_2 \rightarrow B_3$ is proper and twice continuously differentiable on the closed ball. Then f is spherically equivalent to one of the following four mappings:

$$f(z, w) = (z, w, 0)$$
$$f(z, w) = \left(z, zw, w^2\right)$$
$$f(z, w) = \left(z^2, \sqrt{2}zw, w^2\right)$$
$$f(z, w) = \left(z^3, \sqrt{3}zw, w^3\right). \tag{53}$$

Faran proved Theorem 6 assuming three continuous derivatives, but Cima and Suffridge [CS2] reduced the differentiability hypothesis to two continuous derivatives. That only two continuous derivatives are required follows also from Theorem 13 (due to Forstnerič), stated later in this chapter.

We shall see that there are always infinitely many distinct spherical equivalence classes as soon as $N \geq 2n$. This is so even for quadratic monomial mappings. Thus $N = 2n - 1$ is the critical dimension. When $n \geq 2$, and $N = 2n - 1$, there are finitely many distinct classes. From the one ball, according to Theorem I, there are infinitely many classes. The explanation for this is the tensor product. We will not pursue this here, except to say that the simplest nontrivial example is

$$
\begin{aligned}
f(z) &= \left(z_1, \ldots, z_{n-1}, z_1 z_n, \ldots, z_n^2\right) \\
&= E(A, z)(z),
\end{aligned}
\tag{54}
$$

obtained by tensoring the identity with itself on a one-dimensional subspace.

Since the automorphism group of the ball is so large, superficially different mappings can often be spherically equivalent. On the other hand, in this section we will exhibit multidimensional families of polynomial mappings that are mutually inequivalent. We restrict ourselves to the polynomial case for notational ease, but the rational case is not much more difficult. It is not difficult to show [D6] that polynomial mappings preserving the origin are spherically equivalent only if they are unitarily equivalent, so the verification that mappings in our families are inequivalent becomes a matter of linear algebra.

We need some preliminary comments on Hermitian forms. If

$$
Q(z, \bar{z}) = \sum_{|a|=|b|=m} c_{ab} z^a \bar{z}^b
\tag{55}
$$

is the indicated homogeneous real-valued polynomial, then linear algebra or the methods of Chapter 3 show how to write it in the form

$$
Q(z, \bar{z}) = \left\|Q^+(z)\right\|^2 - \left\|Q^-(z)\right\|^2.
\tag{56}
$$

The expressions inside the norms are holomorphic homogeneous polynomials. The necessary and sufficient condition that one can write

$$
Q(z, \bar{z}) = \left\|Q^+(z)\right\|^2
\tag{57}
$$

is simply that the matrix of coefficients (c_{ab}) be positive semidefinite. Simple examples show that this can fail even when $Q(z, \bar{z}) \geq 0 \ \forall z$. We will identify a homogeneous real-valued polynomial of the form (55) with the corresponding Hermitian form on the vector space of homogeneous holomorphic polynomials of degree m. Thus we can say that such a real-valued polynomial is positive semidefinite if and only if it is the norm squared of a holomorphic polynomial mapping.

Using these ideas, we prove the following result which indicates how abundant proper polynomial mappings are in the high codimension case.

PROPOSITION 1
Let

$$p_j : \mathbb{C}^n \to \mathbb{C}^{N_j} \qquad j = 1, \ldots, m-1 \tag{58}$$

be homogeneous vector-valued polynomial mappings of degree j. Suppose that the Hermitian form defined by

$$Q(z, \bar{z}) = ||z||^{2m} - \sum_{j=0}^{m-1} ||p_j(z)||^2 ||z||^{2m-2j}$$

$$= ||H_m||^2 - \sum_{j=0}^{m-1} ||p_j \otimes H_{m-j}||^2 \tag{59}$$

is positive semidefinite. This implies that there is a p_m so that $Q(z, \bar{z}) = ||p_m(z)||^2$. Consider the polynomial mapping p of degree m defined by

$$p = p_0 \oplus p_1 \oplus \cdots \oplus p_m. \tag{60}$$

Then p maps B_n properly to the appropriate B_K.

PROOF Write $p = p_0 + p_1 + \cdots + p_{m-1} + p_m$ for the desired mapping, where p_m is to be determined. We must satisfy the orthogonality relations (36). By decreeing that all the homogeneous components of different orders are orthogonal, we satisfy all the conditions except the condition that

$$\sum_{j=0}^{m} ||p_j||^2 = 1 \qquad \text{on} \qquad ||z||^2 = 1. \tag{61}$$

Homogenizing this equation as before gives the identity (38) that holds on the whole space. Solving this equation for $||p_m||^2$ yields

$$||p_m(z)||^2 = ||z||^{2m} - \sum_{j=0}^{m-1} ||p_j(z)||^2 ||z||^{2m-2j} . \tag{62}$$

To find such a p_m, we need the form defined by the right-hand side of (62) to be positive semidefinite, which is the hypothesis. If this holds, then (62) defines p_m up to a unitary matrix, and finishes the proof. ∎

This simple proposition has many corollaries. One of the simplest arises from choosing

$$p_0 = 0$$
$$p_1 = L. \tag{63}$$

Then there is a quadratic polynomial proper mapping

$$Qz = Lz \oplus \left(\left(\sqrt{I - L^*L} \right) z \otimes z \right) \tag{64}$$

whenever the matrix $I - L^*L$ is positive semidefinite. There are, in particular, infinitely many inequivalent proper quadratic polynomial mappings arising in this fashion. We write these explicitly in dimensions one and two.

Example 3
The process implicit in Proposition 1 yields the following family of mappings from the one ball to the two ball. It is not hard to see that these mappings are spherically inequivalent as the parameter θ varies in the closed interval $[0, \pi/2]$

$$f_\theta(z) = \left(\cos(\theta) z, \sin(\theta) z^2 \right). \tag{65}$$

One interpretation of this result is that the family yields a homotopy between $f_0(z) = z$ and $f_{\pi/2}(z) = z^2$. It is necessary to map into the two ball to have enough room to see this homotopy. ☐

Example 4
To obtain higher dimensional families, we map from the two ball. The process yields a mapping of the form

$$f(z, w) = Az + Bw + Cz^2 + Dzw + Ew^2 \tag{66}$$

to the five ball, where

$$A, B, C, D, E \in \mathbb{C}^5 \tag{67}$$

are vectors that can be chosen as follows:

$$A = (a_1, a_2, 0, 0, 0)$$
$$B = (b_1, b_2, 0, 0, 0)$$
$$C = (0, 0, c, 0, 0)$$
$$D = (0, 0, d_3, d_4, d_5)$$
$$E = (0, 0, 0, 0, e). \tag{68}$$

The vectors satisfy the following conditions:

$$\langle A, B \rangle = - \langle D, E \rangle = - \langle C, D \rangle = \lambda$$

$$\langle A, C \rangle = \langle A, D \rangle = \langle A, E \rangle = 0$$

$$\langle B, C \rangle = \langle B, D \rangle = \langle B, E \rangle = \langle C, E \rangle = 0$$

$$||A||^2 + ||C||^2 = 1$$

$$||B||^2 + ||E||^2 = 1$$

$$||D||^2 = ||C||^2 + ||E||^2. \tag{69}$$

The parameter λ can be nonzero. In coordinates we obtain

$$\left(a_1 z + b_1 w, a_2 z + b_2 w, c z^2 + d_3 z w, d_4 z w, d_5 z w + e w^2 \right) \tag{70}$$

where the last three components are determined up to an element of $U(3)$ by the first two. The first two components can be also subjected to transformation by an element of $U(2)$. Hence the actual number of parameters is three; one can choose them to be

$$\langle A, B \rangle, ||A||^2, ||B||^2. \tag{71}$$

The mapping is given more succinctly by

$$Lz \oplus \left(\left(\sqrt{I - L^* L} \, z \right) \otimes z \right) \tag{72}$$

but there may be some value in expressing it as in (70). The three given parameters determine the linear part up to a unitary transformation, and the linear part determines the quadratic part up to another unitary transformation.

□

It is easy to count the number of parameters in the general case. The following result is useful because it shows already for quadratic polynomials that there are large-dimensional parameter spaces.

PROPOSITION 2
The (spherical equivalence classes of) proper mappings from B_n given by (64) depend on $[n(n-1)]/2$ complex and n real parameters. These parameters are given by the inner products of the column vectors of the linear part. They satisfy the inequalities

$$0 \leq s_j \leq 1$$

$$|t_{jk}| \leq s_j s_k. \tag{73}$$

Two proper mappings of the form (64) are spherically equivalent if and only if the corresponding parameters agree.

PROOF (Sketch). The main point is that proper polynomial mappings that preserve the origin are spherically equivalent only if they are unitarily equivalent [D6]. This assertion can be checked directly here. The relevant parameters are those numbers preserved by unitary transformations; since the quadratic part is determined up to unitary transformations by the linear part, the relevant parameters become those determined by the linear part. The inequalities from (62) follow immediately from the Cauchy–Schwarz inequality and because L must satisfy $L^*L \leq I$. ∎

COROLLARY 3

There are infinitely many spherically inequivalent proper quadratic polynomial mappings $f : B_n \to B_{2n}$.

PROOF The easiest way to see this is to choose the matrix

$$L = \begin{pmatrix} I & 0 \\ 0 & t \end{pmatrix} \tag{74}$$

where the top left entry is the $(n-1) \times (n-1)$ identity matrix. By Proposition 2, each value of $t \in [0,1]$ gives rise to an inequivalent mapping. After a linear transformation, the range can be chosen to be $2n$-dimensional. ∎

Note that the parameter $t \in [0,1]$ gives rise to a homotopy between the mappings

$$f(z) = (z, 0)$$
$$g(z) = \left(z_1, \ldots, z_{n-1}, z_1 z_n, \ldots, z_n^2, 0 \right). \tag{75}$$

Every rational proper mapping between balls can be considered as being obtained by specialization of parameters of a proper map into a ball of perhaps much larger dimension. Thus the space of rational mappings of any bounded degree to a sufficiently large dimensional ball is contractible. These considerations are not difficult, but we will not pursue them here.

We close this section by discussing a generalization of Proposition 2. The question is this. Given a polynomial

$$p : \mathbb{C}^n \to \mathbb{C}^N$$
$$p = p_0 + p_1 + \cdots + p_{m-1} \tag{76}$$

it is natural to ask whether it can be the Taylor polynomial of a proper holomorphic mapping between balls. Let us call such a Taylor polynomial an allowable jet of order $m - 1$. Let us denote by $V(n, N, m-1)$ the vector space of polynomial mappings

$$p : \mathbb{C}^n \to \mathbb{C}^N \tag{77}$$

of degree at most $m - 1$. It is not difficult to prove the following result about allowable jets.

THEOREM 7

For given integers $n, m - 1$ suppose that N is sufficiently large. Then there is an open subset $A \subset V(n, N, m - 1)$ with the following properties:

1. *Each element of A is an allowable jet of order $m-1$ of a proper polynomial mapping between balls of degree m.*

2. *The origin (the trivial jet) is a boundary point of A and is allowable.*

3. *The set A is defined by finitely many polynomial inequalities.*

SKETCH OF THE PROOF   See [D9] for details. One must suppose first that the coefficients of the distinct monomials in a given jet are linearly independent, or else the jet may not be allowable. This is a generic condition and requires that the target dimension be sufficiently large. Next one observes that the trivial jet is allowable, because one can use H_m from Theorem 3. The orthogonality conditions (36) can be all met in case the coefficients are linearly independent, and the last condition in (36) can be met as in Proposition 1. This is a condition on positive semidefiniteness of an Hermitian form. The condition holds at the origin, and the matrix corresponding to $||z||^{2m}$ has minimum eigenvalue equal to unity. Therefore the condition remains true nearby. Furthermore it is determined by finitely many polynomial inequalities.   ∎

There is an alternate approach to the understanding of polynomial proper mappings between balls. Given such a polynomial proper mapping, consider the list of monomials that occur in it. Suppose that these monomials are $c_\alpha z^\alpha$ where the coefficients are vectors. Then the monomial mapping defined by $\sum_\oplus ||c_\alpha|| z^\alpha$, where we take the orthogonal sum, is also a proper mapping between balls. Its range may be higher dimensional. Since the transformation from this mapping back to the original one is linear, it follows that it is sufficient to classify all monomial mappings in order to classify all polynomial ones. For fixed dimensions of the domain and range, finding all monomial examples is an amusing algebraic exercise best done on computer. For $n = 2, N = 3$ there are the examples of Theorem 6. For $n = 2, N = 4$, the author did this by hand. The list appears in [D6,D8], although one example was inadvertently (and incorrectly) dropped from the list. That mapping is $\left(z^2, \sqrt{2}zw, zw^2, w^3\right)$. For fun we now describe how to find all monomial examples given the domain and range dimensions.

Suppose that $f(z) = (\ldots, c_\alpha z^\alpha, \ldots)$ is a proper mapping between balls. Then $\sum |c_\alpha|^2 |z|^{2\alpha} = 1$ on the sphere. Write $x = (x_1, \ldots, x_n) = (|z_1|^2, \ldots, |z_n|^2)$. This condition then becomes $\sum |c_\alpha|^2 x^\alpha = 1$ on the hyperplane $\sum x_j = 1$. Suppose we seek a mapping of degree m. We are asking for a polynomial (in the x variables) with nonnegative coefficients that equals unity on this hyperplane.

In principle, solving this is elementary, but in practice there are many solutions. For example, when $n = 2, N = 5$, there are more than two hundred discrete examples plus examples that depend on a parameter. The maximum degree is 7. We say more about this technique in the next section when we study invariant mappings.

5.2 Invariance under fixed-point–free finite unitary groups

5.2.1 Introduction

We have seen so far that there are many (inequivalent) rational proper mappings between balls. It is natural to ask whether we can find examples with additional properties. One such property is invariance under some subgroup of the unitary group $U(n)$. This question is very natural because odd-dimensional complete connected spaces of constant positive curvature (spherical space forms) are precisely the quotients of odd-dimensional spheres by fixed-point–free finite subgroups of the unitary group [Wo]. Thus (the restriction to the sphere of) an invariant proper mapping induces a map on the spherical space form. It is necessarily a CR mapping. We will determine for which fixed-point–free groups Γ there are proper mappings that are Γ-invariant. As has been common in this book, the one-dimensional case motivates deeper discussion.

Example 5
The proper holomorphic mapping $f(z) = z^m$ from the unit disc to itself is invariant under a cyclic group of order m generated by an mth root of unity. In other words, let G denote the group of one by one matrices generated by ϵ where $\epsilon^m = 1$. If γ is an element of this group, then $f(\gamma z) = f(z)$. Notice that all finite subgroups of the circle $U(1)$ are of this form. Thus every such group occurs as the group of invariants of a proper mapping from the disc to itself.   ☐

Example 6
Let G_m denote now the cyclic group of order m, but represented as n by n matrices by ϵI, where I is the identity and again ϵ is an mth root of unity. Then the proper holomorphic mapping $H_m(z)$ defined by (24) is invariant under G_m.
☐

It is time to begin to sketch the solution to the general problem. We say that a group "occurs" if there is a rational proper holomorphic mapping between balls invariant under it. Whether a group occurs depends both on the group and on its representation.

DEFINITION 1   *A fixed-point–free finite unitary group is a representation*

$$\pi : G \to \mathbf{U}(n) \tag{78}$$

of a finite group G as a group of unitary matrices on $\mathbb{C}^n$, such that 1 is not an eigenvalue of any $\pi(\gamma)$ for $\gamma \in G$ unless $\gamma =$ identity.

We emphasize $\Gamma = \pi(G)$ because the group of matrices is more important than the underlying abstract group. All cyclic groups occur, but not all representations of cyclic groups are possible for rational mappings. If we consider proper mappings that are not smooth at the boundary, then all fixed-point–free finite unitary groups occur. We refer to [Fo1] for a proof of the following result.

THEOREM 8
(Forstnerič). *Let*

$$\pi : G \to \pi(G) = \Gamma \subset \mathbf{U}(n) \tag{79}$$

be a fixed-point–free finite unitary group. Then there is a proper holomorphic mapping f from B_n to some B_N that is invariant under $\Gamma = \pi(G)$. One can assume that f is continuous on the closed ball.

REMARK 5   If one demands that f be smooth on the closed ball, then not all fixed-point–free groups occur. According to another result of Forstnerič (Theorem 13 of this chapter) such mappings would have to be rational. It turns out to be impossible to find rational proper mappings between balls that are invariant under most fixed-point–free groups. Forstnerič discovered some restrictions in [Fo1]. Lichtblau gave additional restrictions in his thesis [L1]. The author and Lichtblau completely answered the question in [DL]. We sketch these results in Section 2.3. Other than Example 6, there is essentially only one other general class of examples, found by the author in [D8]. This is the subject of the next section.   ∎

5.2.2   A class of invariant mappings

Let $\Gamma(2r + 1, 2)$ denote the group of 2 by 2 matrices of the form

$$\begin{pmatrix} \epsilon & 0 \\ 0 & \epsilon^2 \end{pmatrix}^k \tag{80}$$

for $k = 0, 1, \ldots, 2r$ and where ϵ is a primitive root of unity with $\epsilon^{2r+1} = 1$. Then $\Gamma(2r + 1, 2)$ is a fixed-point–free representation of a cyclic group of odd order; this representation differs from the one that occurs for homogeneous polynomials. We write the representation for Example 6 as $\Gamma(p, 1)$ when the degree of homogeneity is p. We have seen in that example that the homogeneous mappings are invariant under this representation of the cyclic group. The content of the next theorem is that there are mappings invariant under the more complicated representation given by (80).

THEOREM 9
For each nonnegative integer r, there is a proper polynomial mapping

$$f : B_2 \rightarrow B_{2+r} \tag{81}$$

invariant under

$$\Gamma(2r + 1, 2). \tag{82}$$

PROOF   We write down the mapping explicitly. Call the variables (z, w). The first step is to list a basis for the algebra of all polynomials invariant under the group. We leave the simple proof that this is a basis to the reader. The basic polynomials are

$$z^{2r+1}, z^{2r-1}w, \ldots, z^{2(r-s)+1}w^s, \ldots, zw^r, w^{2r+1}. \tag{83}$$

As there will be a monomial example, we consider

$$f(z, w) = \left(z^{2r+1}, \ldots, c_s z^{2(r-s)+1}w^s, \ldots, w^{2r+1}\right) \tag{84}$$

and seek to choose the coefficients to make the function map the sphere to the sphere. Computation of $\|f\|^2$ yields

$$1 = \sum_{s=0}^{r} |c_s|^2 |z|^{2(2(r-s)+1)} |w|^{2s} + |w|^{2(2r+1)} \tag{85}$$

on the sphere. The following surprising changes of variables enable us to solve for the coefficients. The first substitution is

$$|z|^2 \rightarrow \frac{1}{t} \qquad |w|^2 \rightarrow \frac{t-1}{t}. \tag{86}$$

Substitution and clearing denominators yields

$$t^{2r+1} = \sum_{s=0}^{r} |c_s|^2 (t(t-1))^s + (t-1)^{2r+1}$$

$$t^{2r+1} - (t-1)^{2r+1} = \sum_{s=0}^{r} |c_s|^2 (t(t-1))^s. \tag{87}$$

The form of (87) suggests the next replacement $t = u + \frac{1}{2}$. This yields

$$\left(u + \frac{1}{2}\right)^{2r+1} - \left(u - \frac{1}{2}\right)^{2r+1} = \sum_{s=0}^{r} |c_s|^2 \left(u^2 - \frac{1}{4}\right)^s. \tag{88}$$

Next we observe that the left side depends only on u^2, so it makes sense to change variables once again, putting $u^2 - 1/4 = x$. After making this substitution and multiplying out the left side, one obtains two equal polynomials in x. Upon equating coefficients, one obtains the formulas for the unknown coefficients:

$$|c_s|^2 = \left(\frac{1}{4}\right)^{r-s} \sum_{k=s}^{r} \binom{2r+1}{2k} \binom{k}{s}. \tag{89}$$

This formula determines the coefficients of the proper mapping up to a unitary transformation. ∎

Theorem 9 has an interesting topological consequence. Recall from topology that the quotient spaces of spheres by groups such as $\Gamma(2r+1,2)$ are examples of Lens spaces.

DEFINITION 2   *The Lens space $L(p,q)$ is the smooth manifold defined by*

$$S^3/\Gamma(p,q). \tag{90}$$

Here $\Gamma(p,q)$ is the group generated by the matrix

$$\begin{pmatrix} \varepsilon & 0 \\ 0 & \varepsilon^q \end{pmatrix} \tag{91}$$

where ε is a primitive pth root of unity, p,q are relatively prime, and we assume without loss of generality that $q < p$. We also usually make the normalization that ε is chosen so that q is minimal.

Since $L(p,q)$ is locally the sphere, it is an example of a CR manifold. Thus it makes sense to ask whether there are nontrivial CR mappings from Lens spaces to other CR manifolds. It turns out that the examples in Corollary 4 are the only ones possible (except for changes in notation).

COROLLARY 4
There are nontrivial smooth CR immersions from the Lens spaces $L(p,1)$ and $L(2r+1,2)$ to sufficiently high dimensional spheres. Here p,r are arbitrary positive integers.

PROOF   We have seen already that the restriction of the homogeneous polynomial mapping H_p induces such a mapping in the first case. In the second case, the restriction of the mapping in Theorem 9 induces an example. These examples are immersions, because their derivatives are injective away from the origin. As the restrictions of holomorphic mappings, they define CR mappings. ∎

The squared coefficients $|c_s|^2$ have many interesting properties. They are always integers. Let us introduce also the parameter r into the notation, and write the squared coefficients as

$$K_{rs} = \left(\frac{1}{4}\right)^{r-s} \sum_{k=s}^{r} \binom{2r+1}{2k} \binom{k}{s}$$

$$= \binom{2r-s}{s-1} + \binom{2r-s+1}{s}. \tag{92}$$

We leave it as an exercise to verify that the simpler second expression holds. There is a nice recurrence relationship that expresses how to get K_{rs} from the various K_{ts} for $t < s$. Those readers who wish to discover this formula can do so by algebraic reasoning or simply by enough staring at the triangle (93) formed by these integers. Those who find numerical puzzles intriguing should attempt to compute the next row. Notice that we have included a one on the far right; this is the coefficient of the $|w|^{2(2r+1)}$ term in (85). Immediately following the triangle we offer the solution to this puzzle.

$$
\begin{array}{ccccccccc}
 & & & & 1 & 1 & & & \\
 & & & 1 & 3 & 1 & & & \\
 & & 1 & 5 & 5 & 1 & & & \\
 & 1 & 7 & 14 & 7 & 1 & & & \\
 1 & 9 & 27 & 30 & 9 & 1 & & & \\
1 & 11 & 44 & 77 & 55 & 11 & 1 & & \\
1 & 13 & 65 & 156 & 182 & 91 & 13 & 1 & \\
1 & 15 & 90 & 275 & 450 & 378 & 140 & 15 & 1
\end{array}
\tag{93}
$$

ANSWER   To find an element (other than the one on the far right and the obvious odd integer preceding it), add up the two numbers directly above the desired slot together with the diagonal ending in the right-hand term. Thus, for example, 156=44+77+27+7+1. These numbers satisfy other recurrences, some shown to the author by Bruce Reznick, who gave an alternate derivation of the formula for them, from which the equality in (92) follows.

REMARK 6   If one analyzes the equations for the coefficients $|c_s|^2$ in a direct manner, then it becomes difficult to verify that the solutions are nonnegative. This difficulty is significant if one considers other groups of this sort. For example, it is fairly easy to verify the following assertion. Suppose that one lists the basic monomials that generate the algebra of polynomials under the group $\Gamma(p,q)$. Then, it is impossible to find solutions to the equations for the absolute valued squared of the coefficients unless the parameter q equals one or two. We will show (by a different method) that there is no $\Gamma(p,q)$-invariant mapping at all if $q \geq 3$. $\blacksquare$

Example 7

From (84) it is possible to construct invariant mappings from the 2-ball to much higher dimensional balls, by taking tensor products. It is less obvious that one can construct an invariant mapping from the 3-ball that is closely related to the special case $\Gamma(7,2)$. This mapping appears in the theses of both Chiappari and Lichtblau; the latter gives a combinatorial derivation. To do so, consider the cyclic group Q of order 7, now represented by the three by three matrices of the form

$$\begin{pmatrix} \epsilon & 0 & 0 \\ 0 & \epsilon^2 & 0 \\ 0 & 0 & \epsilon^4 \end{pmatrix} \tag{94}$$

and suppose that ϵ is a primitive seventh root of unity. Then there is a mapping

$$f : B_3 \to B_{17} \tag{95}$$

that is invariant under Q. To expedite the writing of this mapping, we use the trick described at the end of Section 1.4. The mapping f is monomial, with 17 components. Let us write

$$x = |z_1|^2$$
$$y = |z_2|^2$$
$$u = |z_3|^2 \tag{96}$$

and $f^{\#}(x, y, u) = \|f(z)\|^2$. Since f is a monomial mapping, it is easy to read it off from $f^{\#}$, by taking square roots of each term and replacing the plus signs by commas. The condition that f is proper becomes the condition that

$$f^{\#}(x, y, u) = 1 \quad \text{on} \quad x + y + u = 1 \tag{97}$$

The reader can verify himself, or check on a computer, that the following formula for $f^{\#}$ does the trick:

$$\begin{aligned} f^{\#}(x, y, u) = {}& x^7 + y^7 + u^7 + 14\left(x^3 y^2 + x^2 u^3 + y^3 u^2 + xyu\right) \\ &+ 7\left(x^5 y + xu^5 + y^5 u + xy^3 + x^3 u + yu^3\right) \\ &+ 7\left(xy^2 u^4 + x^2 y^4 u + x^4 yu^2 + x^2 y^2 u^2\right). \end{aligned} \tag{98}$$

The mapping f corresponding to $f^{\#}$ is invariant under the group Q and maps to the ball in 17 dimensions. One can use the minimal invariant polynomial from the next section to discover Example 7. $\square$

5.2.3 Restrictions on the possible groups

In this section we discuss the geometry of invariant mappings. It turns out that proper rational mappings can be invariant only under cyclic groups, and there are severe restrictions on the possible representations. The examples we have seen are essentially the only ones possible. Let us study the homogeneous mappings from the 2-ball in more detail. Recall that

$$\Gamma(m, 1) \tag{99}$$

is the group of 2 by 2 matrices generated by

$$\begin{pmatrix} \epsilon & 0 \\ 0 & \epsilon \end{pmatrix} \tag{100}$$

where $\epsilon^m = 1$. Then we can study quotients of the sphere by this group, obtaining Lens spaces. The homogeneous polynomial mapping of degree m then has the following interpretation. The image of the 2-ball under this mapping is a domain in the complex analytic variety determined by $\mathbb{C}^{m+1}/\Gamma(m, 1)$. This variety has an isolated singularity at the origin; this follows because the group is fixed-point–free. The mapping H_m is then an m-fold covering map on the complement of the origin. In particular, this shows that, for m at least 2, the corresponding variety is not locally Euclidean near the image of 0; if it were, then removing a point would yield a simply connected space. The existence of this m-fold covering shows that this set is not simply connected. The mapping has full rank on the boundary, so its image is a strongly pseudoconvex domain with real analytic boundary. In this case the image embeds explicitly as a ball in $\mathbb{C}^{m+1}$. Since the mapping is a polynomial, it is smooth up to the boundary.

In the general case of a fixed-point–free group, one can try to imitate this. The difference is that the embedding into some complex vector space cannot always be made smooth at the boundary. The reason is that a smooth proper mapping between balls must be rational, but for rational proper mappings between balls, most groups cannot occur. We carry out this program, begun by Forstnerič, in the rest of this section. One step is Theorem 10; this appears in [Li1]. See also [D11].

The elimination of some representations of cyclic groups arises because of a condition on the number of factors of two in an integer. We write $v_2(p)$ for this number. It is harder to eliminate other representations.

THEOREM 10
There is no proper holomorphic mapping $f : B_2 \to B_N$ that is smooth up to the boundary and is also invariant under the group $\Gamma(p, q)$ if

$$v_2(p) > v_2(q - 1). \tag{101}$$

PROOF   According to the result of Forstnerič (Theorem 13 of the next section) a proper holomorphic mapping that is smooth up to the boundary must be a rational function. After composing with an automorphism in the range ball, we may assume that the origin in $\mathbb{C}^2$ is mapped to the origin in $\mathbb{C}^N$. The resulting mapping remains rational and invariant. We may therefore suppose from the start that

$$f = \frac{g}{h} \tag{102}$$

where $h(0) = 1$ and $g(0) = 0$. If a rational mapping is invariant under the group, and the denominator has a nonzero constant term, then we may assume that both the numerator and denominator are also invariant.

As a consequence of the identities (36), we obtain that $\deg(g) > \deg(h)$. As usual we write

$$g(z, w) = \sum_{a,b} c_{ab} z^a w^b$$

$$h(z, w) = \sum_{a,b} d_{ab} z^a w^b. \tag{103}$$

It follows from identity (36) that

$$\sum_{a,b} \left(||c_{ab}||^2 - |d_{ab}|^2 \right) |z|^{2a} |w|^{2b} = 0 \quad \text{on} \quad |z|^2 + |w|^2 = 1. \tag{104}$$

Supposing that $\deg(g) = s$, we substitute $|w|^2 = 1 - |z|^2$ and consider the highest order term. Noting also that $d_{ab} = 0$ for $a + b = s$ reveals that

$$0 = \sum_{a+b=s} \left(||c_{ab}||^2 - |d_{ab}|^2 \right) (-1)^b = \sum_{a+b=s} ||c_{ab}||^2 (-1)^b. \tag{105}$$

Now we exploit the group invariance. Since the group is diagonal, the only monomials that can appear are invariant monomials. Therefore we have the condition that

$$\epsilon^{a+bq} = 1 \Rightarrow$$

$$a + bq = kp \tag{106}$$

for some positive integer k. Combining these two equations reveals that the integers a, b satisfy

$$a + bq = kp$$

$$a + b = s$$

$$s = kp - b(q - 1). \tag{107}$$

If $v_2(p) > v_2(q - 1)$, then p must be even. Since $(p, q) = 1$, $q - 1$ is also even, and therefore so is s. Divide both sides of the last equation in (107) by 2^d, where $d = v_2 (q - 1)$. That equation and the hypothesis (101) imply that

$$b \equiv 2^{-d}s \bmod 2. \tag{108}$$

Since s, the degree of the numerator, is a fixed integer, one sees that each b that occurs in the sum has the same parity. In either case, $(-1)^b$ is independent of b. This contradicts (105), because some coefficient is nonzero. ∎

COROLLARY 5

A smooth (necessarily rational) proper mapping between balls cannot be invariant under any fixed-point–free group that contains a subgroup generated by the matrix

$$\begin{pmatrix} i & 0 \\ 0 & -i \end{pmatrix}. \tag{109}$$

PROOF Since $i^4 = 1$, $v_2 (4) = 2 > v_2 (3 - 1) = 1$. The corollary thus follows from the theorem. ∎

This corollary was proved first by Forstnerič [Fo1]. He began the program to eliminate certain groups as possibilities under which a proper rational mapping between balls could be invariant. We use the phrase "eliminate a group Γ" to mean that we can prove that there is no rational proper mapping between balls that is Γ-invariant. The method used to eliminate groups involves several steps. The classification of fixed-point–free finite unitary groups appears in [Wo]; we do not require all the information there. For our purposes it suffices to divide the groups into three classes. It follows from [Wo] that any fixed-point–free group must lie in one of these classes. The first class, written class I, consists of those groups that contain a quaternionic subgroup. We shall say equivalently that a group is of class I if there is a two-dimensional subspace on which the group has an element that acts as (109). Theorem 10 eliminates these groups. A group is of class II if it has a generator of the form (119) below. A group is of class III if it is cyclic and diagonally generated. We say more about this later. These classes are not disjoint, but they include all the fixed-point–free groups.

We require a generalization of the polarization lemma from Chapter 1 in order to solve the question in general. By polarizing the condition that a proper mapping takes the sphere to the sphere, and assuming the mapping extends holomorphically past, we obtain

$$\langle f(z), f(\bar{\zeta}) \rangle - 1 = h(z, \zeta) \left(\langle z, \bar{\zeta} \rangle - 1 \right). \tag{110}$$

From this we obtain the following simple reflection principle.

COROLLARY 6

(Reflection principle). *If* $f : B_n \to B_N$ *is proper, and holomorphic past the boundary, then*

$$\left\langle f(z), f\left(\frac{z}{\|z\|^2}\right) \right\rangle = 1 \qquad (111)$$

whenever f is defined at both points.

PROOF   Choose $\bar\zeta$ equal to $z/\|z\|^2$ in (110).   ∎

Next we need an invariant version of (110). The usefulness of Definition 3 derives in part from the lemma following it.

DEFINITION 3   *Let* Γ *be a fixed-point–free finite unitary group. The minimal invariant polynomial of* Γ *is the polynomial* $\Upsilon_\Gamma = \Upsilon$ *defined by*

$$\Upsilon(z, \bar z) = -\prod_{\gamma \in \Gamma}\left(1 - \langle \gamma z, z\rangle\right). \qquad (112)$$

LEMMA 3

Suppose that $p : \mathbb{C}^n \to \mathbb{C}^m$ *and* $q : \mathbb{C}^n \to \mathbb{C}^k$ *are holomorphic,* Γ-*invariant maps. That is,* $p(\gamma z) = p(z), q(\gamma z) = q(z)$ *for all group elements* $\gamma \in \Gamma$. *Suppose that*

$$\Phi(z, \bar z) = \|p(z)\|^2 - \|q(z)\|^2 \qquad (113)$$

vanishes on the unit sphere. Then $\Phi(z, \bar z)$ *vanishes on each of the real analytic sets given by*

$$\langle \gamma z, z\rangle - 1 = 0. \qquad (114)$$

In particular, for some real analytic function g we have (locally)

$$\Phi(z, \bar z) = -g(z, \bar z)\prod_{\gamma \in \Gamma}\left(1 - \langle \gamma z, z\rangle\right)$$

$$= g(z, \bar z)\,\Upsilon(z, \bar z). \qquad (115)$$

PROOF   From the definitions, the fact that $\Phi(z, \bar z)$ vanishes on the sphere, and the invariance, one obtains

$$\begin{aligned}
\Phi(z, \bar z) &= \|p(z)\|^2 - \|q(z)\|^2 \\
&= \|p(\gamma z)\|^2 - \|q(\gamma z)\|^2 \\
&= \Phi(\gamma z, \overline{\gamma z}) \\
&= h(\gamma z, \overline{\gamma z})\left(\langle \gamma z, \gamma z\rangle - 1\right).
\end{aligned} \qquad (116)$$

Recalling the antiholomorphicity in the second factor in the inner product, we polarize. We may therefore substitute ζ for $\overline{\gamma z}$, and obtain

$$\Phi(\gamma z, \zeta) = \langle p(\gamma z), p(\overline{\zeta}) \rangle - q(\gamma z)\overline{q}(\zeta)$$
$$= h(\gamma z, \zeta)\left(\langle \gamma z, \overline{\zeta} \rangle - 1\right). \tag{117}$$

Using the invariance (117) becomes

$$\langle p(z), p(\overline{\zeta}) \rangle - q(z)\overline{q}(\zeta) = h(\gamma z, \zeta)\left(\langle \gamma z, \overline{\zeta} \rangle - 1\right). \tag{118}$$

Now resubstitute $\overline{z}$ for ζ in (118) to obtain that $\Phi(z, \overline{z})$ vanishes on each of the sets given by $\langle \gamma z, z \rangle - 1 = 0$. Hence $\Phi(z, \overline{z})$ is divisible by all of the terms $\langle \gamma z, z \rangle - 1$, and (115) follows for an appropriate function g. ∎

It is possible to eliminate noncyclic groups without using these ideas, but the general case seems to require them. That $\Phi(z, \overline{z})$ vanishes along these other sets simplifies things greatly in all cases. We begin by eliminating the noncyclic groups.

THEOREM 11
Suppose that $G \subset U(n)$ is a fixed-point–free finite unitary group. Suppose that $f : B_n \to B_N$ is a rational proper holomorphic mapping that is invariant under G. Then G must be cyclic.

SKETCH OF PROOF   One begins with the classification of the fixed-point–free groups from [Wo]. It is shown there that all such groups that do not contain quaternionic subgroups and are not cyclic must have a generator of the following form. We have called these groups of class II.

$$\begin{pmatrix} 0 & 1 & 0 & 0 & \cdots & 0 \\ 0 & 0 & 1 & 0 & \cdots & 0 \\ 0 & 0 & 0 & 1 & \cdots & 0 \\ 0 & 0 & 0 & 0 & \cdots & 0 \\ & & & \cdots & & \\ \delta & 0 & 0 & 0 & \cdots & 0 \end{pmatrix} \tag{119}$$

Here δ is a primitive root of unity. Suppose that g/h is our alleged rational mapping. In case $n \geq 3$, we put $z_n = 1$, $z_1 = \overline{\delta}$, $z_j = 0$. It then follows that

$$z_2\overline{z_1} + z_3\overline{z_2} + \ldots \delta\overline{z_n} = 1. \tag{120}$$

This means that the point z lies on the set defined by $\langle \gamma z, z \rangle = 1$, where γ is the generator (119). On the other hand, $\|z\|^2 = 1 + |\delta|^2 = 2 > 1$, so this point lies outside the ball. We then have $\|g(z)\|^2 = |h(z)|^2$ by (120). Applying the reflection principle (111), we have the point $w = z/||z||^2$ inside where

$$1 = \left\langle \frac{g(w)}{h(w)}, \frac{g(z)}{h(z)} \right\rangle \le \left\| \frac{g(w)}{h(w)} \right\| \left\| \frac{g(z)}{h(z)} \right\| \le \left\| \frac{g(w)}{h(w)} \right\|. \tag{121}$$

This contradicts the maximum principle. For $n \ge 3$, a generator of the form (119) is therefore not possible. In case $n = 2$, one proceeds as follows. The matrix (119) is then unitarily equivalent to one of the form

$$\begin{pmatrix} \alpha & 0 \\ 0 & -\alpha \end{pmatrix} \tag{122}$$

for some pth root of unity α. We claim that the group can be fixed-point–free only if $p \equiv 0 \bmod (4)$. To see this, first assume p is odd. Raising (122) to the pth power gives a group element, not the identity, with an eigenvalue of one. If $p \equiv 2 \bmod (4)$, then raising (122) to the power $p/2$ has the same effect. Thus we have proved the claim. This means that an appropriate power of (122) is

$$\begin{pmatrix} i & 0 \\ 0 & -i \end{pmatrix}. \tag{123}$$

Therefore there is a subgroup of the type ruled out by Theorem 10. We conclude that all groups of class II are eliminated. Since we have eliminated groups of class I, only groups of class III remain. These are all cyclic. $\blacksquare$

Suppose that Γ is a group of type III. We then see that Γ is generated by a diagonal matrix with eigenvalues that are all powers of a fixed primitive root of unity:

$$\begin{pmatrix} \epsilon^{r_1} & 0 & 0 & 0 \\ 0 & \epsilon^{r_2} & 0 & 0 \\ \cdots & \cdots & \cdots & \cdots \\ 0 & 0 & 0 & \epsilon^{r_n} \end{pmatrix}. \tag{124}$$

The exponents are all relatively prime to p. If an invariant mapping exists, then we can restrict to a two-dimensional subspace, and still obtain an invariant mapping. Thus it is enough to eliminate rational mappings g/h from the two ball that are invariant under $\Gamma(p, r)$.

Thus our plan is to eliminate groups generated by

$$\begin{pmatrix} \epsilon & 0 \\ 0 & \epsilon^r \end{pmatrix}$$

whenever $r \ge 3$, $\epsilon^p = 1$, $r < p$, $(p, r) = 1$. We assume also that r is minimal. The invariant minimal polynomial for $\Gamma(p, r)$ is

$$-\prod_{j=0}^{p-1} \left(1 - \epsilon^j |z|^2 - \epsilon^{jr} |w|^2 \right). \tag{125}$$

Since this polynomial depends only on the squared moduli, we write $x = |z|^2$, $y = |w|^2$ when it is convenient.

Consider the set defined by

$$\epsilon^j x + \epsilon^{jr} y = 1. \tag{126}$$

We seek some j for which there is a solution to (126) satisfying

$$0 \leq x, y$$
$$x + y > 1. \tag{127}$$

If we can find such a point, then we have a point z_0 outside the unit ball where

$$\|g(z_0)\|^2 - |h(z_0)|^2 = 0. \tag{128}$$

By using the reflection principle in Corollary 6, we obtain that

$$\left\| g\left(\frac{z_0}{\|z_0\|^2} \right) \right\|^2 \geq \left| h\left(\frac{z_0}{\|z_0\|^2} \right) \right|^2. \tag{129}$$

This again contradicts the maximum principle, because $\|g\|^2/|h|^2$ must be less than unity on the open unit ball.

To find such a point, it is in some cases sufficient to set $x = y$. In those cases, we need

$$0 < \epsilon^j + \epsilon^{rj} = \frac{1}{x} < 2. \tag{130}$$

We need to ensure that the middle term is real. This is clear if $r = p - 1$. It will be real if

$$jr \equiv -j \mod (p), \tag{131}$$

which holds if

$$j(r + 1) \equiv 0 \mod (p). \tag{132}$$

The possibility that j equals zero is not allowed by (126). As long as $r + 1$ is a zero divisor mod (p), we can find such a j. Using this method, one can rule out groups of the form $\Gamma(p, p - 1)$ when $p \geq 5$.

To handle the general case, a better approach is to set

$$\epsilon^j = e^{-i\theta}. \tag{133}$$

We solve the linear system of equations

$$1 = \cos(\theta)x + \cos(r\theta)y$$
$$0 = \sin(\theta)x + \sin(r\theta)y \tag{134}$$

that arises by expressing (126) in terms of real and imaginary parts. From (134) results the solution

$$x = \frac{-\sin(r\theta)}{\sin(\theta)\cos(r\theta) - \cos(\theta)\sin(r\theta)}$$

$$y = \frac{\sin(\theta)}{\sin(\theta)\cos(r\theta) - \cos(\theta)\sin(r\theta)} \tag{135}$$

as long as the denominator does not vanish. Manipulation of the denominator shows that it equals $-\sin((r-1)\theta)$. Hence we seek to find θ such that

$$\theta = \tfrac{2\pi}{p}k$$

$$\sin(\theta) > 0$$

$$\sin((r-1)\theta) < 0$$

$$\sin(r\theta) < 0 \tag{136}$$

all hold for some integer $k < p/2$. The resulting solution then must satisfy $x + y > 1$ because of the first equation in (134). We state this conclusion as a proposition.

PROPOSITION 3
A proper holomorphic rational mapping between balls cannot be invariant under the group $\Gamma(p,r)$ *if there is a solution to (136).*

It is possible to analyze completely the inequalities (136). First observe that in case $r = 1$ or $r = 2$, there is no solution. Hence we assume that $r \geq 3$, and hence that $p \geq 4$. We require

$$\frac{2\pi}{p} \leq \frac{2\pi}{p}k < \pi$$

$$\pi < \frac{2\pi}{p}k(r-1) < 2\pi$$

$$\pi < \frac{2\pi}{p}kr < 2\pi. \tag{137}$$

Simplifying these inequalities yields

$$1 \leq k < \frac{p}{2}$$

$$\frac{p}{2(r-1)} < k < \frac{p}{r}. \tag{138}$$

Since we are assuming $r > 2$, it is enough to find a positive integer in the second interval. The second interval certainly contains an integer if

$$\frac{p}{r} - \frac{p}{2(r-1)} > 1 \tag{139}$$

from which results

$$p > \frac{2r(r-1)}{r-2} = 2r + \frac{2r}{r-2} . \tag{140}$$

If p is small relative to r, then we proceed as follows. Setting $p = r+1,\ r+2,\ldots$ in turn, we see that the integer unity works for k all the way up to $p = 2(r-1)$. In this case, however, we have $v_2(p) > v_2(r-1)$, so the group is eliminated by Theorem 10. If $p = 2r - 1$, then the condition fails, and there is an invariant mapping. In this case, however, a change of notation (obtained by squaring the generator) shows that $\Gamma(2r-1,r) = \Gamma(2r-1,2)$; we analyzed these groups in Theorem 9. The case $p = 2r$ is not possible, because relative primeness is violated. We now consider the cases $p = 2r+1, 2r+2, \ldots$. It is convenient to handle the cases $r = 3$ and $r = 4$ separately. We have the chart

$$\begin{array}{ll} \Gamma(7,3) & k = 2 \\ \Gamma(8,3) & v_2(8) > v_2(3-1) \\ \Gamma(9,3) & \text{rel.prime violated} \\ \Gamma(10,3) & k = 3 \\ \Gamma(11,3) & k = 3 \\ \Gamma(12,3) & \text{rel.prime violated} \end{array} \tag{141}$$

where the second entry gives the explanation for elimination and k satisfies (138). By (140), the remaining cases (for larger p) when $r = 3$ are eliminated, because

$$p > 2r + \frac{2r}{r-2} = 12.$$

Thus we have eliminated $\Gamma(p,3)$ for all p.

For $r = 4$ we need to check only the four cases in (142):

$$\begin{array}{ll} \Gamma(9,4) & k = 2 \\ \Gamma(10,4) & k = 2 \\ \Gamma(11,4) & k = 2 \\ \Gamma(12,4) & \text{rel.prime violated.} \end{array} \tag{142}$$

For $p > 12$ we eliminate $\Gamma(p,4)$ by (140).

We may now assume that $r \geq 5$. According to (140), we need only to check the cases $p = 2r + 1,\ 2r + 2,\ 2r + 3$. For each of these three cases we always have that

$$\frac{p}{2r-2} < 2 < \frac{p}{r} . \tag{143}$$

Therefore $\Gamma(2r+1,r)$, $\Gamma(2r+2,r)$, and $\Gamma(2r+3,r)$ are always ruled out when $r \geq 5$.

Therefore there is always a positive integer solution to (138) unless we are in the six cases listed below:

$$r = 1$$
$$r = 2$$
$$p = 2(r - 1)$$
$$p = 2r - 1$$
$$p = 2r$$
$$p = 8, \quad r = 3. \tag{144}$$

The first, second, and fourth have invariant mappings. The fourth is really the same as the second. The fifth does not matter because relative primeness is violated. We eliminated the third in Theorem 10. The special case $\Gamma(8, 3)$ does not occur also because of Theorem 10.

We have proved the following theorem.

THEOREM 12
Let $\Gamma(p, r)$ be the group generated by

$$\begin{pmatrix} \epsilon & 0 \\ 0 & \epsilon^r \end{pmatrix}$$

where $\epsilon^p = 1$, $(p, r) = 1$, and r is chosen minimally. Suppose that there is a proper holomorphic rational mapping from B_2 to some ball that is $\Gamma(p, r)$-invariant. Then necessarily either $r = 1$ or $r = 2$. In each of these cases, there is a polynomial example.

A summary of these results is now appropriate, as we have completely solved the problem. Groups of class I are eliminated by Theorem 10. Groups of class II are eliminated by Theorem 11. Groups of class III can occur, but only in the following three cases. We have eliminated all other representations. Thus a group that occurs must be cyclic, and only these representations are possible.

Case 1. The group is generated by

$$(\epsilon I_n) \tag{145}$$

with no restriction on the root of unity. See Theorem 3 for an example.

Case 2. The group is generated by

$$\begin{pmatrix} \epsilon I_k & 0 \\ 0 & \epsilon^2 I_j \end{pmatrix} \tag{146}$$

where ϵ is an odd root of unity. See Theorem 9 for an example.

Case 3. The group is generated by

$$
\begin{pmatrix}
\epsilon I_{k_1} & 0 & 0 \\
0 & \epsilon^2 I_{k_2} & 0 \\
0 & 0 & \epsilon^4 I_{k_3}
\end{pmatrix}
\tag{147}
$$

where $\epsilon^7 = 1$. See Example 7.

In each case the mapping f satisfies $\|f\|^2 - 1 = \Upsilon$, where Υ is the minimal invariant polynomial of the group.

The following corollary relates this theorem to our earlier work.

COROLLARY 7

There is no smooth nonconstant CR mapping from the spherical space form S^{2n-1}/Γ to any sphere, unless the group Γ is one of the examples (145), (146), or (147). For every Γ there exist continuous CR mappings from S^{2n-1}/Γ to some sphere.

PROOF If such a smooth CR mapping existed, then it could be extended to be an invariant proper holomorphic mapping between balls. This mapping would be smooth at the boundary, hence rational, so the first part of the corollary follows from the above summary. The second statement follows from Theorem 8, by considering the restriction of the continuous invariant mapping to the sphere.
∎

It is also of interest to consider other finite unitary groups. It is not difficult to verify [Fo3] that only fixed-point–free groups occur for smooth mappings. More precisely, suppose that $f : B_n \to \Omega$ is a proper holomorphic mapping. Suppose that $b\Omega$ is smooth, f is smooth up to the boundary, and f is invariant under a group G. Then G must be fixed-point–free. On the other hand, if one allows domains with nonsmooth boundaries, then many interesting examples arise, including reflection groups. We refer to [Ru3] for general results, but give one nice example.

Example 8
Put

$$
f(z, w) = \left(z^2 - w^2, z^4 + w^4 \right). \tag{148}
$$

Then $f : B_2 \to \Omega$ is a proper holomorphic mapping with multiplicity eight from the ball to a domain Ω. It is not hard to check that it is invariant under the group D_4 of symmetries of the square. More precisely, consider the matrix group generated by

$$A = \begin{pmatrix} 1 & 0 \\ 0 & -1 \end{pmatrix}$$

$$B = \begin{pmatrix} 0 & i \\ i & 0 \end{pmatrix}. \tag{149}$$

Then these generators satisfy the relations:

$$A^2 = I$$

$$B^4 = I$$

$$AB = B^3 A \tag{150}$$

and hence define a group isomorphic to the group of the square. It is clear that (148) is invariant under these matrices. ∎

5.3 Boundary behavior

5.3.1 Forstnerič's theorem on the rationality of smooth proper mappings

We have seen that all proper holomorphic mappings from a ball to itself are rational. This result no longer holds when the dimension of the range ball exceeds the dimension of the domain ball. It is easy to see that there are proper mappings $f : B_1 \to B_2$ that are not rational. To do so, choose a pair of real-valued, continuous functions on the unit circle that satisfy

$$e^{2u_1} + e^{2u_2} = 1. \tag{151}$$

We extend these functions to be harmonic in the unit disc. We may find harmonic conjugate functions so that

$$f = \left(e^{(u_1 + iv_1)}, e^{(u_2 + iv_2)} \right) \tag{152}$$

is a holomorphic mapping from the unit disc to $\mathbb{C}^2$. Computing

$$\|f\|^2 = e^{2u_1} + e^{2u_2} = 1 \tag{153}$$

on the circle, and using the maximum principle, we see that $f : B_1 \to B_2$ is a proper holomorphic mapping between balls. Since we could have chosen one of the functions u_1 almost arbitrarily, the resulting proper mapping need not be rational. This result also suggests the hopelessness of trying to classify the proper mappings between balls, unless we make additional hypotheses about the mappings. In [Fo2], Forstnerič proved that a sufficiently differentiable proper holomorphic mapping between balls is necessarily rational. He proved the following general theorem:

THEOREM 13

(Forstnerič). *Suppose that $N > n > 1$,*

$$\Omega_1 \subset\subset \mathbb{C}^n$$
$$\Omega_2 \subset\subset \mathbb{C}^N \tag{154}$$

are strongly pseudoconvex domains with real analytic boundaries, and $f : \Omega_1 \to \Omega_2$ is a proper holomorphic mapping between them. If also $f \in C^\infty\left(\overline{\Omega_1}\right)$, then it extends holomorphically past a dense open subset of the boundary. If the second domain is a ball, then the result holds when f is assumed to be only of class C^{N-n+1}. Furthermore, if both domains are balls, then f is a rational function.

REMARK 7 In case f is a rational proper holomorphic mapping between balls, it extends holomorphically past the sphere. Thus there cannot be a point of indeterminacy on the sphere. This result was proved independently by Pinchuk [Pi] and by Cima and Suffridge [CS3]. An extension of this result was proved also by Chiappari [Ch]. Forstnerič also obtains a crude bound on the degree of the defining polynomials. The examples in Section 2.2 suggest what may be the actual bound, in the special case that $2 = n \leq N$. The conjecture is that the degree is no larger than $2N - 3$; note that this degree is achieved for the invariant mappings (84). ∎

The proof of Theorem 13 involves certain varieties associated with the proper mapping. In case the mapping extends holomorphically past the boundary, these varieties have a pleasing definition. The idea of Forstnerič's proof is to define these varieties by considering only a finite Taylor development at the boundary. Once one has their definition, there remains work to do to show that they are rational. To gain some feeling for this, we consider them for mappings that are known already to be holomorphic past the sphere. The polarization trick of Chapter 1 is the main idea!

5.3.2 Varieties associated with proper mappings

DEFINITION 4 *Suppose that $f : (\mathbb{C}^n, 0) \to (\mathbb{C}^N, 0)$ is a germ of a holomorphic mapping. Suppose further that M, M' are real analytic hypersurfaces containing the respective origins, and $f : M \to M'$. Finally, suppose that r, r' are defining functions for these hypersurfaces. Then the X-variety associated* with the mapping f is the germ at the origin of a complex analytic subvariety of $\mathbb{C}^{n+N}$ defined by

$$X_f = \{(w,\zeta) : r'(f(z),\zeta) = 0 \text{ when } r(z,w) = 0\}. \tag{155}$$

We omit the simple verification that these varieties are independent of the choice of defining function. It is important to notice that the X-varieties extend the graph of f; this is the following lemma.

LEMMA 4
Under the hypotheses of Definition 4, the point $(w, f(w))$ lies in X_f.

PROOF   Since $f : M \to M'$, there is an equality of the form

$$r'\left(f(z), \overline{f(z)}\right) = h(z,\bar{z})\, r(z,\bar{z}) \tag{156}$$

for some local real analytic function h. According to Proposition 1.1 on polarization, we can substitute w for $\bar{z}$ and the result follows.   ∎

It is important to observe that the X-varieties are usually much larger than the graph. It is also possible for them to correspond exactly to the graph, for nontrivial mappings. We compute some examples of these varieties for proper mappings between balls. In the case of balls, the equations are that

$$\langle f(z), \zeta \rangle = 1 \quad \text{when} \quad \langle z, w \rangle = 1. \tag{157}$$

These equations mean geometrically that f maps each complex hyperplane not through the origin to a corresponding complex hyperplane with normal vector ζ. In the language of Section 2, we could say that the minimum invariant polynomial of the trivial group is $||z||^2 - 1$.

Example 7
For the mapping H_m, the X-variety equals the graph.   ∎

PROOF   Recall Theorem 3. After a unitary transformation, we may assume that the mapping H_m is defined by

$$H_m(z) = (\ldots, c_a z^a, \ldots) \tag{158}$$

where we are using the notation of that theorem. First we consider the polarized condition on the domain sphere. Using the multinomial expansion, this condition becomes

$$\langle z, w \rangle = 1 \Rightarrow$$
$$\langle z, w \rangle^m = 1 \Rightarrow$$
$$\sum_{|a|=m} \binom{m}{a} (z\overline{w})^a = 1. \tag{159}$$

Now we write the polarized condition on the range sphere as

$$\sum_{|a|=m} c_a z^a \overline{\zeta_a} = 1. \tag{160}$$

These expressions are equal for all z. Equating power series coefficients shows that

$$\binom{m}{a} \overline{w^a} = c_a \overline{\zeta_a}. \tag{161}$$

Using the values of c_a that arise from knowing that $\zeta = H_m(w)$ works, we not only see that $\zeta_a = c_a w^a$, but also that

$$|c_a|^2 = \binom{m}{a}.$$

This gives, therefore, a second proof of the essential uniqueness of the homogeneous proper polynomial mapping in addition to verifying that $\zeta = f(w)$. Statement (161) implies that the X-variety equals the graph of the mapping. ∎

Example 8

For the simplest nontrivial proper polynomial mapping

$$f(z_1, z_2) = \left(z_1^2, z_1 z_2, z_2 \right), \tag{162}$$

the X-variety is larger than the graph. It is as follows:

$$\left(w_1, w_2, w_1^2, w_1 w_2, w_2 \right) \cup \left(w_1, 0, w_1^2, -w_1 t, t \right). \tag{163}$$

To verify this, we suppose that $z_1 \overline{w_1} + z_2 \overline{w_2} = 1$ and plug into the equation

$$z_1^2 \overline{\zeta_1} + z_1 z_2 \overline{\zeta_2} + z_2 \overline{\zeta_3} = 1. \tag{164}$$

If w_2 does not vanish, then one eliminates z_2 and obtains three equations for $(\zeta_1, \zeta_2, \zeta_3)$ by equating coefficients of $1, z_1, z_1^2$. This procedure yields the graph. If, however, w_2 does vanish, then the resulting system has rank 2, and we obtain an arbitrary parameter in the solution set. In addition to points in the graph, one obtains also points of the form

$$\left(w_1, 0, w_1^2, -w_1 t, t \right) \tag{165}$$

for every complex number t. It is clear that these points lie in the graph only when $t = 0$. □

For the multidimensional families of proper polynomial mappings between balls, the X-varieties become exceedingly complicated. They need not even be pure dimensional. Their importance at present is that Forstnerič uses them in his proof of Theorem 13. Let us indicate how the X-variety bears on the rationality of the mapping. Consider the equation

$$\langle f(z), \zeta \rangle = 1 \tag{166}$$

on the sphere. We may differentiate (166) in the direction of the tangential (1,0) vector fields L_j from Chapter 3. By differentiating enough times, one obtains on the sphere a system of linear equations for ζ. If we assume that there is a unique solution to these equations, that is, when the X-variety equals the graph, then the solution must be $f(z)$. On the other hand, the solution to such a system of linear equations can be given by Cramer's rule as a ratio of two determinants. The resulting expression will then be a rational function but will involve both $z, \bar{z}$. By considering each variable separately, one can replace (on the sphere) with unbarred terms the barred terms that appear. One obtains a formula that extends f beyond the sphere.

We illustrate this in the trivial one-dimensional case. Equation (166) becomes

$$f(z)\overline{f(z)} = 1 \quad \text{on} \quad z\bar{z} = 1. \tag{167}$$

It is not necessary to differentiate to obtain enough equations! We get

$$\overline{f(z)} = \frac{1}{f(z)} = \frac{1}{f\left(\frac{1}{\bar{z}}\right)} \quad \text{on} \quad z\bar{z} = 1 \tag{168}$$

and hence we obtain the formula that

$$f(z) = \frac{1}{f\left(\frac{1}{\bar{z}}\right)} \tag{169}$$

for the extension of f beyond the disc. The resulting function is meromorphic and has no essential singularities (including at infinity), so it must be rational.

The proof of Theorem 13 uses similar ideas, but is more elaborate in part because one must consider the case where the system of equations has infinitely many solutions. It seems to be an interesting question to decide what role the X-varieties play in the classification of the proper mappings between balls.

5.3.3 Nonsmooth proper mappings

In this section we do little more than state some results. The basic question is as follows. Suppose that $f : \Omega_1 \subset\subset \mathbb{C}^n \to \Omega_2 \subset\subset \mathbb{C}^N$ is a proper holomorphic mapping, and suppose also that the boundaries of each domain are smooth manifolds. Under what circumstances must the mapping extend smoothly to the boundaries? It was proved by Kellogg in 1912 that in case both domains are one-dimensional, the proper mapping extends. It also has been known for a long time that the result does not hold for proper mappings from $B_1 \to B_2$. See the discussion concerning (151). It remains an open problem whether there are any counterexamples in the equidimensional case, although Barrett [Bar1] has constructed counterexamples for the regularity of biholomorphic mappings for domains in complex manifolds that are not Stein. It is a relatively recent discovery that there are nonsmooth proper mappings between balls whenever the dimension of the target ball is strictly larger than the dimension of the domain ball. This result was obtained independently by Løw [Lw] and Forstnerič [Fo4] when the codimension was large, and was later generalized to the codimension-one case by, also independently, Hakim [Ha] and Dor [Dr]. The reader should consult the survey [Fo3].

Another unknown aspect of the theory is how smooth a proper mapping must be before it must be rational. No one has ever exhibited a proper mapping between balls, where the domain ball is in $\mathbb{C}^n$ for $n > 1$, that is continuously differentiable without being smooth and rational. Also there is an interesting connection between smoothness and invariance under unitary groups. The content of Theorem 8 is that, for any fixed-point–free finite unitary group, one can find a proper mapping between balls invariant under this group. Since not all such groups can occur for rational mappings, one sees that invariant mappings are in general not smooth at the boundary. On the other hand, it is easy to verify that unitary groups that are not fixed-point–free cannot give rise to smooth invariant mappings between balls.

Positive codimension gives one great latitude. The recent theorems of Dor and Hakim yield proper holomorphic mappings from any smoothly bounded strongly pseudoconvex domain into a ball in one larger dimension that is continuous at the boundary. If one wishes to make the mapping smooth, then this is not possible [Fo4]. Most strongly pseudoconvex domains with even real analytic boundary cannot be properly embedded into a ball in any finite dimension such that the embedding is smooth at the boundary. Lempert [Le1] has shown recently that one can achieve embeddings into balls in Hilbert spaces. One of his results amounts to saying that every real analytic strongly pseudoconvex (necessarily bounded) domain can be defined by $||f||^2 < 1$ for a Hilbert space valued mapping f.

6

Geometry of the $\bar\partial$-Neumann Problem

6.1 Introduction to the problem

A fundamental theorem in the subject of function theory of several complex variables is the solution of the Levi problem, establishing the identity of domains of holomorphy and pseudoconvex domains. A proof includes the solution of the Cauchy–Riemann equations on such domains. Theorem 1 gives the three most important equivalent properties about domains. See [Hm,Kr,Ra] for definitions, other equivalences, and complete proofs.

THEOREM 1
The following are equivalent for a domain $\Omega \subset \mathbb{C}^n$.

1. *Ω is a domain of holomorphy.*

2. *Ω is a pseudoconvex domain.*

3. *For all $q \geq 1$ and for all smooth $(0,q)$ forms α such that $\bar\partial\alpha = 0$, there is a smooth $(0, q-1)$ form u such that $\bar\partial u = \alpha$. In the language of sheaf cohomology,*

$$H^q(\Omega, O) = 0, \qquad q \geq 1. \tag{1}$$

The solvability of the Cauchy–Riemann equations in the smooth category yields essentially the solution to the Levi problem. Solving the Cauchy–Riemann equations in other categories is natural both as a generalization of the Levi problem and as part of the theory of partial differential equations. The literature along these lines is enormous. In this book we are interested in boundary behavior of mappings. Suppose that the domain in question has smooth boundary. The study of boundary behavior of holomorphic functions leads us to the study of the boundary behavior of the Cauchy–Riemann equations. Can one solve the inhomogeneous Cauchy–Riemann equations $\bar\partial u = \alpha$, so that the solution u is smooth on the closure of the domain, given that $\bar\partial\alpha = 0$ and α is smooth there?

Consider the case of $(0,1)$ forms. Suppose that $\alpha \in C_{0,1}^{\infty}\left(\overline{\Omega}\right)$ is a smooth $(0,1)$ form on the closed domain, and $\overline{\partial}\alpha = 0$. Suppose that u is a smooth solution to the equation

$$\overline{\partial}u = \alpha. \tag{2}$$

On any pseudoconvex domain it is possible [Hm] to find a holomorphic function that cannot be smoothly extended to a given boundary point. If f is such a function, then $u + f$ is also a solution of (2), because

$$\overline{\partial}\left(u + f\right) = \overline{\partial}u = \alpha. \tag{3}$$

Note that the solution $u + f$ is not smooth on the closed domain. Since not all solutions are smooth, we seek some particular solution that is. It is plausible from these remarks that the solution orthogonal to the holomorphic functions should have optimal regularity properties. Although this result is not known, the aim of the $\overline{\partial}$-Neumann problem is to study the particular solution of the Cauchy–Riemann equations that is orthogonal to the null space of $\overline{\partial}$.

The set up follows. See [FK,K1,K3,K4] for details. We wish to solve the overdetermined system of first-order partial differential equations

$$\overline{\partial}u = g \quad \text{given} \quad \overline{\partial}g = 0 \tag{4}$$

where g is a differential form of type $(0,q)$ on a domain Ω in $\mathbb{C}^n$. We suppose that Ω is smoothly bounded and carries a Hermitian metric. One usually assumes that this is the Euclidean metric; the metric used in this chapter differs from the Euclidean metric only by factors of two. Since the problem has infinitely many solutions, the idea is to construct one particular solution, namely the unique solution orthogonal to the null space of $\overline{\partial}$ on $(0,q)$ forms. Thus the solution should be in the range of the adjoint. These considerations of orthogonality and adjoints require that one work in L^2 spaces.

We denote by $L_{0,q}^2\left(\Omega\right)$ the square integrable $(0,q)$ forms. If $\phi = \sum_{|J|=q}\phi_J d\overline{z}^J$ is such a form, then we define

$$\|\phi\|^2 = \int_{\Omega}\sum_J|\phi_J|^2\,dV. \tag{5}$$

In the Euclidean metric there would a factor of 2^{-q} in front of the integral. This arises because the 1-form $d\overline{z}^j$ has squared length equal to two in the Euclidean metric at each point of the complex cotangent space. It is often useful in the subject to consider weight functions; this means that one replaces (5) with

$$\|\phi\|_\psi^2 = \int_{\Omega}\sum_J|\phi_J|^2\,e^{-\psi}\,dV \tag{6}$$

where ψ is chosen according to the demands of the problem. We allude to this method later. Unless stated otherwise, we work always with the norm defined by (5).

By defining $\overline{\partial}$ in the sense of distributions, taking its closed maximal extension, and defining the adjoint $\overline{\partial}^*$ of the closed densely defined operator $\overline{\partial}$ as usual, one obtains the diagram

$$L^2_{0,q-1}\left(\Omega\right) \overset{\overline{\partial}}{\underset{\overline{\partial}^*}{\rightleftarrows}} L^2_{0,q}\left(\Omega\right) \overset{\overline{\partial}}{\underset{\overline{\partial}^*}{\rightleftarrows}} L^2_{0,q+1}\left(\Omega\right). \tag{7}$$

For a densely defined operator T we denote its domain by $D(T)$ and its null space by $\mathrm{Ker}(T)$. The domain of $\overline{\partial}$ is

$$D\left(\overline{\partial}\right) = \left\{g \in L^2_{0,q} : \overline{\partial}g \in L^2_{0,q+1}\right\}. \tag{8}$$

Thus a square integrable form g is in $D(T)$ if the distribution (weak) derivatives of each component function with respect to each $\bar{z}_j$ are themselves square integrable. The domain of $\overline{\partial}^*$ is defined as usual by

$$D\left(\overline{\partial}^*\right) = \left\{g \in L^2_{0,q} : f \rightarrow \langle \overline{\partial}f, g \rangle \text{ is cts. on } L^2_{0,q-1}\right\}. \tag{9}$$

The operator $\square$ is defined by

$$\square f = \left(\overline{\partial}\,\overline{\partial}^* + \overline{\partial}^*\overline{\partial}\right) f \tag{10}$$

where the domain of $\square$ is

$$D(\square) = \left\{f : f \in D\left(\overline{\partial}\right), \overline{\partial}f \in D\left(\overline{\partial}^*\right), f \in D\left(\overline{\partial}^*\right), \overline{\partial}^*f \in D\left(\overline{\partial}\right)\right\}. \tag{11}$$

The conditions that a form be in the domain of the adjoint operator are boundary conditions that arise from integration by parts. This is where the geometry of the boundary enters the picture.

To see these things, we first review some advanced calculus. Suppose that r is a defining function for Ω and its gradient has norm one on the boundary. Then the gradient is the exterior unit normal vector field. Letting dV and dS denote the volume and surface area forms, we recall the formula for integration by parts (the divergence theorem):

$$\int_\Omega \frac{\partial f}{\partial x^j} \, dV = \int_{b\Omega} f\frac{\partial r}{\partial x^j} \, dS. \tag{12}$$

Here $\partial/\partial x^j$ is a coordinate vector field. We can now express the boundary conditions for the $\overline{\partial}$-Neumann problem in terms of r. For example, a smooth $(0,1)$ form $\phi = \sum \phi_j d\bar{z}^j$ is in the domain of $\overline{\partial}^*$ precisely when

$$\sum_{j=1}^n \phi_j \frac{\partial r}{\partial z_j} = 0 \quad \text{on} \quad \{r = 0\}. \tag{13}$$

To see this, we have

$$(\overline{\partial} f, \phi) = \sum_{j=1}^{n} \int_{\Omega} \frac{\partial f}{\partial \overline{z}^j} \overline{\phi_j} \, dV$$

$$= -\sum_{j=1}^{n} \int_{\Omega} f \overline{\frac{\partial \phi}{\partial z^j}} \, dV + \sum_{j=1}^{n} \int_{b\Omega} f \overline{\phi_j \frac{\partial r}{\partial z^j}} \, dS. \tag{14}$$

In order that the continuity condition in (9) hold, the boundary integral in (14) must vanish for every f; this requires that $\sum \phi_j (\partial r / \partial z_j) = 0$ on $\{r = 0\}$. Similar formulas hold for forms of all degrees.

The $\overline{\partial}$-Neumann problem is to establish existence and regularity for the system of equations

$$\Box f = \left(\overline{\partial} \, \overline{\partial}^* + \overline{\partial}^* \overline{\partial} \right) f = g \tag{15}$$

for a $(0,q)$ form g such that $g \perp \mathrm{Ker}\,(\Box)$. If there is a solution, then there is a unique solution that is orthogonal to $\mathrm{Ker}\,(\Box)$. The solution is usually denoted by Ng. It makes sense then to define the $\overline{\partial}$-Neumann operator by

$$N = \Box^{-1} \quad \text{on} \quad (\mathrm{Ker}\,(\Box))^{\perp}$$

$$N = 0 \quad \text{on} \quad \mathrm{Ker}\,(\Box). \tag{16}$$

The $\overline{\partial}$-Neumann problem is a boundary value problem that is not elliptic. The operator $\Box$ is elliptic, as it is essentially the Laplacian, but the boundary conditions are not. Nevertheless, Kohn was able to solve this problem through estimates that are called subelliptic estimates. One important feature of a subelliptic estimate is the local regularity property for a solution to the Cauchy–Riemann equations. It follows from the subelliptic estimate (34) introduced in Definition 1 that the solution $u = Ng$ to $\Box u = g$ is smooth in any open set where g is. We next verify that the unique solution to the Cauchy–Riemann equations

$$\overline{\partial} f = g \tag{17}$$

orthogonal to the null space of $\Box$ is given by

$$\overline{\partial}_* N g. \tag{18}$$

This particular solution is consequently smooth wherever g is smooth, assuming that a subelliptic estimate holds.

Let us amplify the connection between the $\overline{\partial}$ and $\overline{\partial}$-Neumann problems. A $(0,q)$ form g has the decomposition

$$g = Hg + \Box Ng = Hg + \left(\overline{\partial} \, \overline{\partial}^* + \overline{\partial}^* \overline{\partial} \right) Ng. \tag{19}$$

Here the operator H is the orthogonal projection onto $\mathrm{Ker}\,(\Box)$. Assuming that

$$Hg = \overline{\partial} g = 0, \tag{20}$$

we apply $\bar{\partial}$ to (19), and obtain

$$0 = \bar{\partial}g = \bar{\partial}\bar{\partial}\bar{\partial}^*\bar{\partial}Ng. \tag{21}$$

Taking inner products,

$$0 = \left\langle \bar{\partial}\bar{\partial}^*\bar{\partial}Ng, \bar{\partial}Ng \right\rangle \Rightarrow$$

$$0 = \left\| \bar{\partial}^*\bar{\partial}Ng \right\|^2 \Rightarrow$$

$$0 = \bar{\partial}^*\bar{\partial}Ng \tag{22}$$

and putting (22) back into (19) gives us the result that

$$g = \bar{\partial}\left(\bar{\partial}^* Ng\right). \tag{23}$$

The particular solution of the $\bar{\partial}$ equation indicated in (23) is called the Kohn solution, canonical solution, or the $\bar{\partial}$-Neumann solution to the Cauchy–Riemann equations.

6.2   Existence and regularity results on the $\bar{\partial}$ and $\bar{\partial}$-Neumann problems

In this section we describe without proofs some of the results of Kohn [K1,K2, K3,K4]. The reader should be aware that there is a vast literature on estimates and regularity results for $\bar{\partial}$ and related operators. There are many methods, including the use of explicit integral solution operators in some cases. See [Ra] for a careful treatment of this method, and see the references there for other techniques. As of 1991, only the $\bar{\partial}$-Neumann approach has yielded general results for domains of finite type in dimensions greater than two.

The existence theory in L^2 includes the following result.

THEOREM 2
Suppose that $\Omega \subset\subset \mathbb{C}^n$ is a smoothly bounded pseudoconvex domain. Then the $\bar{\partial}$-Neumann problem is solvable (in L^2) on $(0,q)$ forms for all q, and the null space $\mathrm{Ker}\,(\Box)$ is trivial on $(0,q)$ forms as long as $q > 0$.

The regularity theory includes the following basic result.

THEOREM 3
Suppose that $\Omega \subset \mathbb{C}^n$ is a pseudoconvex domain with smooth boundary, and that $\alpha \in C^{\infty}_{0,1}\left(\overline{\Omega}\right)$ is a smooth $(0,1)$ form with $\bar{\partial}\alpha = 0$. Then there is a smooth solution $u \in C^{\infty}\left(\overline{\Omega}\right)$ to the equation $\bar{\partial}u = \alpha$.

Theorem 3 gives a global regularity result; it is not known whether the $\bar{\partial}$-Neumann solution is always smooth. For domains of finite type [Ca1], or for domains with plurisubharmonic defining functions [BS2], the $\bar{\partial}$-Neumann solution is smooth on the closed domain. Theorem 3 answers the question posed in the introduction to this chapter. We are also interested in questions of local regularity; if α is smooth on a particular subset of the closed domain, is there a solution to $\bar{\partial}u = \alpha$ that is also smooth there? When subelliptic estimates (see Definition 1) hold, such a local regularity result applies. Furthermore, in terms of Sobolev norms, there is a gain in the degree of differentiability. See [FK] for the fundamentals of the Sobolev theory of derivatives. For integer values of s, H_s denotes the Hilbert space completion of the space of functions whose derivatives up to order s are square integrable. We will need to study fractional derivatives. To see where this is headed, we state a fundamental theorem of Kohn on subellipticity.

THEOREM 4

Suppose that $\Omega \subset \mathbb{C}^n$ is a pseudoconvex domain with smooth boundary. Assume that $0 \in b\Omega$ and there is a subelliptic estimate of order ϵ on $(0, q)$ forms at 0. Suppose that U is an open neighborhood of 0, and that $\alpha \in L^2_{0,q}(\Omega) \cap C^\infty(U)$. Then the $\bar{\partial}$-Neumann solution $\bar{\partial}^* N\alpha$ is also smooth on U. In terms of the Sobolev norms $H_s = H_s(U)$, we have

$$\alpha \in H_s \Rightarrow N\alpha \in H_{s+2\epsilon}$$
$$\alpha \in H_s \Rightarrow \bar{\partial}^* N\alpha \in H_{s+\epsilon}. \tag{24}$$

It also follows that, on $(0,1)$ forms, the Bergman projection $P : L^2(\Omega) \to O(\Omega) \cap L^2(\Omega)$ maps each H_s to itself and hence preserves smooth functions.

The result on the Bergman projection is significant, because it enables one to apply the methods of Bell to show that proper and therefore biholomorphic mappings have smooth extensions to the boundary in quite general situations. When these extension results hold, the numbers we computed in Chapter 4 become biholomorphic invariants of the domain itself. Furthermore they transform in a natural way under proper mappings. We discuss these considerations in Chapter 7.

6.3 Subellipticity

6.3.1 Remarks on tangential pseudodifferential operators

The estimates in (24) are subelliptic estimates when $\epsilon < 1$, because the solution is smoother than the data by an amount less than the order of the operator. To measure smoothness for derivatives of order less than one, it is necessary to use the theory of pseudodifferential operators. In this section we give a brief review of tangential pseudodifferential operators that will enable us to understand such fractional derivatives. See [FK] for more information.

Let $\Omega \subset\subset \mathbf{C}^n$ be a domain with smooth boundary, and let $x \in b\Omega$. Suppose U is a neighborhood of x. We call a local C^∞ coordinate system (defined on U) a system of boundary coordinates if one of the coordinates is a local defining function for the boundary. We write $t = (t_1, \ldots, t_{2n-1})$ for $t \in \mathbf{R}^{2n-1}$ so that $(t, r) \in \mathbf{R}^{2n-1} \times \mathbf{R}$ form local coordinates. We denote the dual variable to t by τ and write $t\tau = \sum_{j=1}^{2n-1} t_j \tau_j$ for notational ease. (This notation will be used only in this section.) Then one has the tangential Fourier transform, defined on $C_0^\infty (U \cap \bar{\Omega})$ by

$$\hat{f}(\tau, r) = \frac{1}{(2\pi)^{\frac{2n-1}{2}}} \int_{R^{2n-1}} e^{-it\tau} f(t, r) \, dt. \tag{25}$$

One can then define tangential pseudodifferential operators in the usual manner by

$$\Lambda^s f(t, r) = \frac{1}{(2\pi)^{\frac{2n-1}{2}}} \int_{R^{2n-1}} e^{it\tau} \hat{f}(\tau, r) \left(1 + |\tau|^2\right)^{\frac{s}{2}} d\tau. \tag{26}$$

These Λ^s operators are used to define the tangential Sobolev norms. The tangential Sobolev norm of order s is defined on all smooth, compactly supported functions by integrating out the various norms on $\mathbf{R}^{2n-1}$:

$$|||f|||_s^2 = \int_{-\infty}^{0} \int_{R^{2n-1}} |\Lambda^s f(t, r)|^2 \, dt \, dr. \tag{27}$$

More generally, we define an algebra of tangential pseudodifferential operators by allowing classical symbols. Thus P is a tangential pseudodifferential operator of order s if the action of P on the smooth compactly supported functions is given by

$$Pu(t, r) = \int_{R^{2n-1}} e^{it\tau} p(t, r, \tau) \hat{u}(\tau, r) \, d\tau. \tag{28}$$

Here the function p, called the symbol of P, is smooth in all its variables and, on every compact subset satisfies the usual estimates:

$$\left| D^\alpha D_\tau^\beta p(t, r, \tau) \right| \le C_{\alpha\beta} \left(1 + |\tau|^{s - |\beta|}\right). \tag{29}$$

The simplest examples are the Λ^s and differential operators of order s where all derivatives are taken in the tangential (t) variables.

We assume without proof the following facts about the algebra of tangential pseudodifferential operators. To clarify the role of the order, we use a superscript to denote the order of the operator. We write P^* for the formal adjoint operator and $[P, Q]$ for the commutator of two tangential pseudodifferential operators. The notation T^k is an abbreviation for "some tangential pseudodifferential operator of order k."

THEOREM 5

(Formal Properties of Pseudodifferential Operators).

1. *If P^m denotes a tangential pseudodifferential operator of order m, then for each s there is a constant c_s such that*

$$|||P^m u|||_s \leq c_s |||u|||_{s+m} \,. \tag{30}$$

2. *If P^m, Q^s denote tangential pseudodifferential operators of the indicated orders, then their commutator is a tangential pseudodifferential operator of order $m + s - 1$:*

$$\left[P^m, Q^s\right] = T^{m+s-1}. \tag{31}$$

3. *Orders add under composition:*

$$P^m Q^s = T^{m+s}. \tag{32}$$

4. *Let $\overline{P}$ denote the operator whose symbol is obtained by conjugation. Then*

$$(P^m)^* - \overline{(P^m)} = T^{m-1}. \tag{33}$$

The proofs of Propositions 1 through 5 concerning subelliptic multipliers require these properties.

6.3.2 Definition of subelliptic estimates

It is now possible to give the definition of a subelliptic estimate. Let us write $D_{0,q}(U)$ for the space of forms of type $(0, q)$ that are in the domain of $\overline{\partial}^*$ and whose coefficients are in $C_o^\infty(U \cap \overline{\Omega})$. The formula $|||\phi|||_\epsilon^2 = \sum_{|J|=q} |||\phi_J|||^2$ defines the squared tangential Sobolev norm of order ϵ of a form $\phi = \sum_{|J|=q} \phi_J d\overline{z}^J \in D_{0,q}(U)$. As usual the sums are taken over strictly increasing multi-indices.

DEFINITION 1 *Suppose that $\Omega \subset\subset \mathbb{C}^n$ is a domain with smooth boundary, and $x \in \overline{\Omega}$. The $\overline{\partial}$-Neumann problem satisfies a subelliptic estimate on $(0, q)$*

forms at x if there is a neighborhood U of x and positive constants c, ϵ such that

$$|||\phi|||_\epsilon^2 \leq c \left(\left|\left|\bar{\partial}\phi\right|\right|^2 + \left|\left|\bar{\partial}^* \phi\right|\right|^2 + ||\phi||^2 \right)$$

$$\phi \in D_{0,q}\,(U)\,. \tag{34}$$

It is usual to denote the quadratic form on the right-hand side of (34) by $Q\,(\phi, \phi)$. In Lemma 1 we compute this expression explicitly in the case of (0,1) forms, and we study this case in detail in this chapter.

It is important to note that the estimate (34), on forms of all degrees, is easy to prove when $x \in \Omega$. We show this (for (0,1) forms) in Lemma 2. Since the $\bar{\partial}$-Neumann problem is elliptic in the interior, the estimate holds for $\epsilon = 1$. The interesting problem lies therefore at the boundary. We will concentrate on subelliptic estimates for $(0, 1)$ forms. Let us compute a completely explicit understandable expression for the quadratic form Q. Suppose that

$$\phi = \sum_{i=1}^{n} \phi_i d\bar{z}^i \tag{35}$$

is a smooth (0,1) form and that it is in the domain $D_{0,1}\,(U)$ of $\bar{\partial}^*$. Recall from (14) that the definition of the adjoint and an integration by parts (the divergence theorem) reveal that $\sum r_{z_i}\phi_i = 0$ on the boundary. It follows that

$$\left(\sum_{i=1}^{n} \phi_i \frac{\partial}{\partial z_i} \right) \tag{36}$$

is a tangential derivative, and hence

$$\overline{\sum_{j=1}^{n} \phi_j \frac{\partial}{\partial z^j}} \left(\sum_{i=1}^{n} \phi_i r_{z_i} \right) = 0 \tag{37}$$

on the boundary. A consequence of (37) is that, on the boundary,

$$\sum_{i,j=1}^{n} \overline{\phi_j}\,(\phi_i)_{\bar{z}_j}\, r_{z_i} = -\sum_{i,j=1}^{n} r_{z_i \bar{z}_j} \phi_i \overline{\phi_j}. \tag{38}$$

The calculation of the adjoint shows also that

$$\bar{\partial}^* \phi = -\sum_{i=1}^{n} (\phi_i)_{z_i}\,. \tag{39}$$

We now perform integration by parts on the definition of the squared norm

$||\bar{\partial}\phi||^2$. Recalling the alternating property of the exterior product, and displaying some skill at index gymnastics, one proceeds as follows:

$$||\bar{\partial}\phi||^2 = \sum_{1\le i<j}^{n} \int_{\Omega} \left|(\phi_i)_{\bar{z}_j} - (\phi_j)_{\bar{z}_i}\right|^2 dV$$

$$= \sum_{i,j=1}^{n} \int_{\Omega} \left|(\phi_i)_{\bar{z}_j}\right|^2 dV - \sum_{i,j=1}^{n} \int_{\Omega} (\phi_i)_{\bar{z}_j} \overline{(\phi_j)_{\bar{z}_i}} \, dV$$

$$= \sum_{i,j=1}^{n} \int_{\Omega} \left|(\phi_i)_{\bar{z}_j}\right|^2 dV + \sum_{i,j=1}^{n} \int_{\Omega} (\phi_i)_{z_i \bar{z}_j} \overline{\phi_j} dV - \sum_{i,j=1}^{n} \int_{b\Omega} (\phi_i)_{\bar{z}_j} r_{z_i} \overline{\phi_j} \, dS$$

$$= \sum_{i,j=1}^{n} \int_{\Omega} \left|(\phi_i)_{\bar{z}_j}\right|^2 dV - \left|\left|\bar{\partial}^* \phi\right|\right|^2 + \sum_{i,j=1}^{n} \int_{b\Omega} r_{z_i \bar{z}_j} \phi_i \overline{\phi_j} \, dS. \tag{40}$$

The third line is the result of integration by parts and the divergence theorem (12). The last line results from another integration by parts on the middle term of the previous line. The resulting new boundary term vanishes because of (13), so that the middle term becomes

$$-\sum_{i,j=1}^{n} \int_{\Omega} (\phi_i)_{z_i} \overline{(\phi_j)}_{z_j} \, dV = -\left|\left|\bar{\partial}^* \phi\right|\right|^2 \tag{41}$$

by using formula (39) for the adjoint. We can rewrite the third term because (38) holds on the boundary. Putting this together shows that

LEMMA 1
The quadratic form Q on $D_{0,1}$ satisfies

$$Q(\phi,\phi) = \sum_{i,j=1}^{n} \int_{\Omega} \left|(\phi_i)_{\bar{z}_j}\right|^2 dV + \sum_{i,j=1}^{n} \int_{b\Omega} r_{z_i \bar{z}_j} \phi_i \overline{\phi_j} \, dS + \int_{\Omega} \sum_{i=1}^{n} |\phi_i|^2 \, dV$$

$$= ||\phi||_{\bar{z}}^2 + \int_{b\Omega} \lambda(\phi,\phi) \, dS + ||\phi||^2. \tag{42}$$

Note that (42) reveals an asymmetry between the barred and unbarred derivatives. The second line consists of convenient abbreviations; the first term $||\phi||_{\bar{z}}$ consists of the barred derivatives. The second term is the integral of the Levi form, and hence is nonnegative on a pseudoconvex domain. It is thus plausible that the closer this form is to being positive definite, the better chance there is to have an estimate.

The first simple application of Lemma 1 is that an elliptic estimate holds in the interior.

LEMMA 2
If $x \in \Omega$, then the estimate (34) holds with $\epsilon = 1$.

PROOF First we have

$$|||\phi|||_1^2 \leq ||\phi||_1^2 = c \sum_{i,j=1}^n \int_\Omega \left|\frac{\partial \phi_i}{\partial z_j}\right|^2 + \left|\frac{\partial \phi_i}{\partial \bar{z}_j}\right|^2 \, dV + ||\phi||^2 \tag{43}$$

It follows immediately from two integrations by parts that

$$\int_\Omega \left|\frac{\partial \phi_i}{\partial z_j}\right|^2 dV = \int_\Omega \left|\frac{\partial \phi_i}{\partial \bar{z}_j}\right|^2 dV \tag{44}$$

if the form ϕ is supported in the interior. Hence, given an interior point, we can choose an open subset of Ω containing this point as the set U in Definition 1. Then

$$|||\phi|||_1^2 \leq ||\phi||_1^2 = c \sum_{i,j=1}^n \int_\Omega \left|\frac{\partial \phi_i}{\partial z_j}\right|^2 + \left|\frac{\partial \phi_i}{\partial \bar{z}_j}\right|^2 \, dV + ||\phi||^2$$

$$= 2c \, ||\phi||_{\bar{z}}^2 + ||\phi||^2$$

$$\leq C \, Q(\phi, \phi) \tag{45}$$

by Lemma 1. Thus the (elliptic) estimate holds in the interior. ∎

We will continue studying this quadratic form after introducing the concept of subelliptic multiplier. First we discuss the connection between subelliptic estimates and the geometry of the boundary.

6.3.3 The relationship to points of finite type

The original motivation for studying "points of finite type" was an attempt to find necessary and sufficient conditions for the subelliptic estimates. After deep work from the mid-fifties through the seventies, Kohn (1979) obtained a sufficient condition for subellipticity on domains with real analytic boundary. By using geometric work of Diederich and Fornaess, this condition is equivalent to the nonexistence of complex analytic varieties in the boundary. The proof in the smooth case came later. Recall from Chapter 4 that a point is of finite type if there is a bound on the maximum order of contact of one-dimensional varieties with the boundary at that point; i.e., $\Delta^1(b\Omega, p) < \infty$. Catlin [C1,C2,C3] established that finite type is a necessary and sufficient condition for subellipticity on pseudoconvex domains. He proved the necessity result by 1983 and the sufficiency result by 1986.

THEOREM 6

(Catlin). *Suppose Ω is a smoothly bounded pseudoconvex domain in $\mathbb{C}^n$ and p is a boundary point. Then the $\bar{\partial}$-Neumann problem satisfies a subelliptic estimate on $(0,1)$ forms at p if and only if p is a point of finite type.*

The sufficiency of the condition is a particularly deep result; as yet no simplifications in the proof have appeared. Let us discuss briefly one aspect of the necessity result. Suppose that the estimate is known to hold for some ϵ. Then it must be true that

$$\epsilon \leq \frac{1}{\Delta^1(b\Omega, p)} \ . \tag{46}$$

This result appeared in [C1]. There is very little known about the largest possible value of the parameter ϵ. The author wrote down a simple example that shows that the value of ϵ is not in general the reciprocal of the order of contact. The ideas have appeared already in the discussion of semicontinuity in Example 4.4.

Example 1

Put

$$r(z, \bar{z}) = 2 \operatorname{Re}(z_3) + \left| z_1^2 - z_2 z_3 \right|^2 + \left| z_2^2 \right|^2 \tag{47}$$

and let Ω be a bounded pseudoconvex domain for which r is a local defining function near the origin. (The zero set of this particular r is unbounded.) In Example 4.4 we computed the order of contact:

$$\Delta^1(M, 0) = 4$$

$$\Delta^1(M, p) = 8 \quad \text{if} \quad p = (0, 0, ia) \ . \tag{48}$$

Since the function

$$p \to \Delta^1(M, p) \tag{49}$$

is not upper semicontinuous near the origin, its reciprocal cannot be lower semicontinuous. It follows immediately from Definition 1 that the same value of the parameter ϵ that works at a point must work near there as well. Thus the value of the largest possible ϵ cannot be the reciprocal of the order of contact. In this particular example, it is possible to prove that the value of ϵ must be $1/8$. ☐

The author believes the following conjecture should be within reach.

Conjecture. Suppose that $\Omega \subset\subset \mathbb{C}^n$ is a smoothly bounded pseudoconvex domain and a subelliptic estimate holds at the point $p \in b\Omega$. Then the estimate holds for

$$\epsilon = \frac{1}{B^1(M,p)} \; . \tag{50}$$

For particular domains the value of ϵ may be larger than the right side of (50). Notice that

$$\frac{1}{B^1(M,p)} \leq \frac{1}{\Delta^1(M,p)} \; .$$

The conjecture implies that the supremum of the set of values of ϵ for which the estimate holds always lies between these values.

Example 2
Let

$$r(z,\bar{z}) = 2 \operatorname{Re}(z_n) + \sum_{j=1}^{n-1} |z_j|^{2m_j} \tag{51}$$

be a local defining function near the origin. At the origin, it is known from the diagonalizability of the Levi form in a neighborhood that the value of ϵ can be chosen to be

$$\epsilon = \frac{1}{2\max(m_i)} = \frac{1}{\Delta^1(M,0)} \; . \tag{52}$$

The reciprocal of the multiplicity is

$$\frac{1}{B^1(M,p)} = \frac{1}{2\prod m_j} \; .$$

For domains defined locally by

$$r = 2 \operatorname{Re}(z_n) + \sum_{j=1}^{n-1} |z_j + z_n g_j(z)|^{2m_j} \, , \tag{53}$$

however, it is not generally possible to take ϵ equal to the reciprocal of the one-type. According to the conjecture, the largest value would be at least as large as the reciprocal of the multiplicity. $\square$

6.4   Subelliptic multipliers

6.4.1   Introduction

In 1979 Kohn [K3] developed the theory of subelliptic multipliers. He invented an interesting algorithmic procedure for computing certain ideals; these ideals, at least in the real analytic case, govern both whether there is a complex analytic variety in the boundary and whether there is a subelliptic estimate. In this section, we give an introduction to the method of Kohn, including proofs, and we discuss a holomorphic analogue of his procedure. This analogue amounts to computing these ideals on domains defined locally by

$$2 \, \text{Re} \left(z_{n+1} \right) + ||f(z)||^2 . \tag{54}$$

We continue to restrict our consideration to the case of (0,1) forms; this is the most important case and is easier only in the sense that the notation is easier to follow. Kohn considers the collection of germs of functions g for which one can estimate the ϵ norm of $g\phi$ in terms of $Q(\phi, \phi)$; the open neighborhood U, the constant c, and the parameter ϵ are all allowed to depend on g. Such functions are known as subelliptic multipliers. More precisely, Kohn made the following definition.

DEFINITION 2 *Let Ω be a smoothly bounded pseudoconvex domain in $\mathbb{C}^n$. Let x be a point in $\overline{\Omega}$, and let C_x^∞ denote the ring of germs of smooth functions there. An element $g \in C_x^\infty$ is called a subelliptic multiplier (on (0,1) forms) if there is a neighborhood U, and positive constants c, ϵ such that*

$$|||g\phi|||_\epsilon^2 \le c \left(||\overline{\partial}\phi||^2 + ||\overline{\partial}^* \phi||^2 + ||\phi||^2 \right)$$
$$\phi \in D_{0,1}(U). \tag{55}$$

The collection of subelliptic multipliers turns out to be a radical ideal. Let us elaborate the notion of radical of an ideal in the ring C_x^∞. The reader should consult [M] for more on these matters. Let J be an ideal in C_x^∞. The radical of J, written $\text{rad}_R(J)$, and sometimes called "the real radical" of J, is the collection of germs $g \in C_x^\infty$ such that there is an integer N and an element $f \in J$ for which

$$|g|^N \le |f| . \tag{56}$$

The term "real radical" emphasizes that the underlying ring consists of (germs of) functions depending on both z and $\overline{z}$. In the real analytic category, we have the same definition (56) of real radical. The theory there resembles that for the ring of holomorphic germs O, but it is considerably more difficult. Comparison with the notion of radical for a ring such as O shows that we have an inequality rather than an equality. A simple example pinpoints why one cannot demand equality in this context. Consider the real function $|z|^2$ in the complex plane. It vanishes precisely at the origin. There are functions vanishing there, for which no power is divisible by $|z|^2$. Take, for example, $\text{Re}(z)$. It is obvious that

$$(\text{Re}(z))^2 \le |z|^2 . \tag{57}$$

To preserve some relationship between ideals and varieties, in the real analytic category, one must therefore allow this broader sense of radicals. The

Lojasiewicz inequality [M,N] then becomes a precise analogue of the Nullstellensatz. This inequality can be stated as follows.

THEOREM 7
A (germ of a) real analytic function f vanishes on the zeroes of an ideal J generated by real analytic functions if and only if

$$f \in \text{rad}_R (J). \tag{58}$$

Thus the ideal of the variety of an ideal J is the real radical of J. This correspondence is crucial in comprehending the geometry of subelliptic multipliers.

6.4.2 Properties of the ideal of subelliptic multipliers

We now establish the fundamental properties of subelliptic multipliers. The collection of all such multipliers defines a real radical ideal. The main point is that a subelliptic estimate holds on $(0,1)$ forms if and only if the ideal of subelliptic multipliers is the full ring of germs of smooth functions. For points in the interior of the domain, this follows from the interior ellipticity of the $\bar{\partial}$-Neumann problem; according to Lemma 2 we may choose ϵ equal to unity. For boundary points one sees easily (see Proposition 2) that the ideal of subelliptic multipliers is not the zero ideal, since the defining function is a subelliptic multiplier with ϵ also equal to unity. Given one multiplier, Kohn gives an algorithmic procedure involving derivatives, determinants, and radicals that determines other multipliers, for which the value of ϵ is typically smaller. The determinant of the Levi form is the first example of a multiplier. The idea is then to obtain a nonvanishing constant after finitely many applications of this procedure.

To begin the discussion, we use the properties of tangential pseudodifferential operators to prove the following propositions of Kohn. The reader should be aware that the value of the constant may change each time it occurs. We also use the following standard trick.

REMARK 1 Suppose we wish to estimate an expression of the form xy. For any positive δ we can write

$$|xy| \leq \frac{\delta}{2}x^2 + \frac{1}{2\delta}y^2 \tag{59}$$

since

$$0 \leq \left(\sqrt{\delta}x - \frac{1}{\sqrt{\delta}}y \right)^2.$$

This is advantageous when we need a small coefficient in front of x^2 but we can allow an arbitrarily large constant in front of y^2. We often refer to $\delta/2$ as a small constant, $1/(2\delta)$ as a large constant, and write (sc) and (lc) for them. ∎

PROPOSITION 1

Let x be a boundary point of Ω. Then the collection of subelliptic multipliers J_x on (0,1) forms is a radical ideal. In particular,

$$g \in J, \quad |f|^N \leq |g| \Rightarrow f \in J. \tag{60}$$

PROOF First we prove that J_x is an ideal. To see this, note that

$$
\begin{aligned}
|||gu|||_\epsilon^2 &= \int |\Lambda^\epsilon (gu)|^2 \\
&= \int |g\Lambda^\epsilon u + P^{\epsilon-1} u|^2 \\
&\leq 2 \int |g|^2 |\Lambda^\epsilon u|^2 + c |||u|||_{\epsilon-1}^2 \\
&\leq c |||u|||_\epsilon^2 + c |||u|||_{\epsilon-1}^2 \\
&\leq c |||u|||_\epsilon^2 .
\end{aligned}
\tag{61}
$$

In (61) $P^{\epsilon-1}$ is the commutator $[\Lambda^\epsilon, g]$. Replacing g by hf, we see that hf is a subelliptic multiplier whenever f is. Thus J_x is closed under multiplication by elements of C_x^∞. The simple verification that the sum of two multipliers is a multiplier will be left to the reader. To verify that the ideal of subelliptic multipliers is a radical ideal, we establish two inequalities. Remain aware that the letter c denotes a positive constant whose value need not be the same for each occurrence. These inequalities, combined with the fact that $\|\phi\|^2 \leq Q(\phi, \phi)$, then imply the desired result. Note that we are assuming that the parameter ϵ is less than one.

$$|||f\phi|||_\epsilon \leq |||g\phi|||_\epsilon + c|||\phi|||_{\epsilon-1} \quad \text{if} \quad |f| \leq |g| \tag{62}$$

$$|||g\phi|||_\epsilon^2 \leq k |||g^m \phi|||_{m\epsilon}^2 + c \|\phi\|^2 \qquad m\epsilon \leq 1. \tag{63}$$

Here k is a positive constant that we may often take to be unity. To prove inequality (62), begin with the definition of the tangential norm. Then move (using Theorem 5) the zeroth-order operator given by multiplication by f, past the Λ^ϵ operator, and use the Schwarz inequality:

$$
\begin{aligned}
|||f\phi|||_\epsilon^2 &= (\Lambda^\epsilon f\phi, \Lambda^\epsilon f\phi) \\
&= ([\Lambda^\epsilon, f] \phi, \Lambda^\epsilon f\phi) + (f\Lambda^\epsilon \phi, \Lambda^\epsilon f\phi) \\
&= (P^{\epsilon-1}\phi, \Lambda^\epsilon f\phi) + (f\Lambda^\epsilon \phi, \Lambda^\epsilon f\phi) \\
&\leq \|P^{\epsilon-1}\phi\| \|\Lambda^\epsilon f\phi\| + \|f\Lambda^\epsilon \phi\| \|\Lambda^\epsilon f\phi\| .
\end{aligned}
\tag{64}
$$

Dividing by $|||f\phi|||_\epsilon$, and because $P^{\epsilon-1}$ is of order $\epsilon - 1$, we obtain the estimate

$$|||f\phi|||_\epsilon \leq c |||\phi|||_{\epsilon-1} + \|f\Lambda^\epsilon \phi\| . \tag{65}$$

The second term on the right depends only on $|f|$ rather than f. This allows us to use the hypothesis that $|f| \leq |g|$ and thereby obtain

$$||| f\phi |||_\epsilon \leq c \, ||| \phi |||_{\epsilon-1} + || f\Lambda^\epsilon \phi ||$$
$$\leq c \, ||| \phi |||_{\epsilon-1} + || g\Lambda^\epsilon \phi ||. \tag{66}$$

Now the same reasoning reveals that

$$|| g\Lambda^\epsilon \phi || \leq || \Lambda^\epsilon g\phi || + c \, ||| \phi |||_{\epsilon-1}. \tag{67}$$

Putting (66) and (67) together establishes the desired result.

To verify the second inequality (63) one performs induction on m, with slight differences according to whether it is even or odd. We work out in agonizing detail the cases where m equals two or three, and leave the details of the induction to the reader. Suppose m equals two:

$$||| g\phi |||_\epsilon^2 = (\Lambda^\epsilon g\phi, \Lambda^\epsilon g\phi) = (\Lambda^{2\epsilon} g\phi, g\phi)$$
$$= (g\Lambda^{2\epsilon}\phi, g\phi) + (P^{2\epsilon-1}\phi, g\phi)$$
$$= \left(|g|^2 \Lambda^{2\epsilon}\phi, \phi \right) + (P^{2\epsilon-1}\phi, g\phi)$$
$$\leq \left| \left| |g|^2 \Lambda^{2\epsilon}\phi \right| \right| \, ||\phi|| + c \, ||| \phi |||_{2\epsilon-1} \, ||\phi||$$
$$= || g^2 \Lambda^{2\epsilon}\phi || \, ||\phi|| + c \, ||| \phi |||_{2\epsilon-1} \, ||\phi||$$
$$= || \Lambda^{2\epsilon} g^2\phi + P^{2\epsilon-1}\phi || \, ||\phi|| + c \, ||| \phi |||_{2\epsilon-1} \, ||\phi||$$
$$\leq ||| g^2\phi |||_{2\epsilon} \, ||\phi|| + c \, ||| \phi |||_{2\epsilon-1} \, ||\phi||$$
$$\leq (sc) \, ||| g^2\phi |||_{2\epsilon}^2 + (lc) \, ||\phi||^2 + c \, ||\phi|| \, ||| \phi |||_{2\epsilon-1}$$
$$\leq (sc) \, ||| g^2\phi |||_{2\epsilon}^2 + C \, ||\phi||. \tag{68}$$

We use the self adjointness of Λ in the first line, then move it past g in the second. In the fifth line we eliminate the absolute value signs because we have an ordinary norm. Note that we also could have used the inequality (62) to eliminate the absolute value signs on $|g|^2$. In the penultimate line we use Remark 1. This device is not really necessary for the proof. By using it we can achieve an arbitrarily small positive constant in front of the $|| g^2\phi ||_\epsilon^2$ term in (63). In the last line we need $2\epsilon - 1 \leq 0$ to finish the proof.

To do the case $m = 3$ requires more of the same and uses the result for $m = 2$. Given the interpolation inequality

$$||| g^2\phi |||_{2\epsilon}^2 \leq c_1 \, ||| g^3\phi |||_{3\epsilon} \, ||| g\phi |||_\epsilon + c \, ||\phi||^2 \tag{69}$$

things are easy. Namely,

$$\begin{aligned}
|||g\phi|||_\epsilon^2 &\le k\,|||g^2\phi|||_{2\epsilon}^2 + c\,||\phi||^2 \\
&\le kc_1\,|||g^3\phi|||_{3\epsilon}\,|||g\phi|||_\epsilon + c\,||\phi||^2 \\
&\le \frac{kc_1}{2}\,|||g^3\phi|||_{3\epsilon}^2 + \frac{kc_1}{2}\,|||g\phi|||_\epsilon^2 + c\,||\phi||^2.
\end{aligned} \tag{70}$$

Assume that $kc_1 < 2$. After subtracting the middle term from each side and multiplying by $2/(2 - kc_1)$, we obtain

$$|||g\phi|||_\epsilon^2 \le \frac{kc_1}{2 - kc_1}\,|||g^3\phi|||_{3\epsilon}^2 + c\,||\phi||^2. \tag{71}$$

If we know the value of the constant c_1, then we can choose k arbitrarily small and again make the coefficient in front as small as we wish. It remains to prove the interpolation estimate (69). To do so, proceed as follows:

$$\begin{aligned}
|||g^2\phi|||_{2\epsilon}^2 &= \left(\Lambda^{2\epsilon}g^2\phi, \Lambda^{2\epsilon}g^2\phi\right) \\
&= \left(\Lambda^{3\epsilon}g^2\phi, \Lambda^\epsilon g^2\phi\right) \\
&= \left(\Lambda^{3\epsilon}g^2\phi, g\Lambda^\epsilon g\phi\right) + \left(\Lambda^{3\epsilon}g^2\phi, P^{\epsilon-1}\phi\right) \\
&= \left(\bar{g}\Lambda^{3\epsilon}g^2\phi, \Lambda^\epsilon g\phi\right) + \left(\Lambda^{2\epsilon}g^2\phi, P^{2\epsilon-1}\phi\right) \\
&\le \left(\bar{g}\Lambda^{3\epsilon}g^2\phi, \Lambda^\epsilon g\phi\right) + (sc)\,|||g^2\phi|||_{2\epsilon}^2 + (lc)\,|||\phi|||_{2\epsilon-1}^2 \\
&\le \left(\bar{g}\Lambda^{2\epsilon}g^2\phi, \Lambda^{2\epsilon}g\phi\right) + (sc)\,|||g^2\phi|||_{2\epsilon}^2 + c\,||\phi||^2 \\
&= \left(\Lambda^\epsilon\bar{g}\Lambda^{2\epsilon}g^2\phi, \Lambda^\epsilon g\phi\right) + (sc)\,|||g^2\phi|||_{2\epsilon}^2 + c\,||\phi||^2.
\end{aligned} \tag{72}$$

We needed $2\epsilon - 1 \le 0$. After subtracting the middle term and multiplying through again, we obtain

$$|||g^2\phi|||_{2\epsilon}^2 \le \frac{1}{1-(sc)}\,||\Lambda^\epsilon\bar{g}\Lambda^{2\epsilon}g^2\phi||\,|||g\phi|||_\epsilon + c\,||\phi||^2. \tag{73}$$

The first term can now be estimated as follows.

$$\begin{aligned}
||\Lambda^\epsilon\bar{g}\Lambda^{2\epsilon}g^2\phi|| &= ||\bar{g}g^2\Lambda^{3\epsilon}\phi + P^{\epsilon-1}\phi|| \\
&\le ||g^3\Lambda^{3\epsilon}\phi|| + c\,|||\phi|||_{3\epsilon-1} \\
&\le |||g^3\phi|||_{3\epsilon} + c\,|||\phi|||_{3\epsilon-1}.
\end{aligned} \tag{74}$$

Using $3\epsilon - 1 \le 0$ we estimate the last term. Putting (74) into (73) and using the result for $m = 2$ with a small coefficient, establishes

$$\left|\left|\left|g^2\phi\right|\right|\right|^2_{2\epsilon} \le \frac{1}{1-(sc)} \left|\left|\left|g^3\phi\right|\right|\right|_{3\epsilon} \left|\left|\left|g\phi\right|\right|\right| + \frac{1}{2\,(1-(sc))} \left|\left|\left|g\phi\right|\right|\right|^2_{\epsilon} + c\left|\left|\phi\right|\right|^2$$

$$\le \frac{1}{1-(sc)} \left|\left|\left|g^3\phi\right|\right|\right|_{3\epsilon} \left|\left|\left|g\phi\right|\right|\right| + (sc_2)\left|\left|\left|g^2\phi\right|\right|\right|^2_{\epsilon} + c\left|\left|\phi\right|\right|^2. \quad (75)$$

We subtract again and multiply through. At last we obtain the interpolation inequality with the constant

$$c_1 = \frac{1}{(1-(sc))\,(1-(sc_2))}$$

which can be made arbitrarily close to unity from above. This finally finishes the proof of (63) for $m = 3$. The general case is virtually the same, where the interpolation inequality used is as in Remark 2.

That the ideal of subelliptic multipliers is a radical ideal follows immediately from (63) and the definition of real radical. ∎

REMARK 2   The interpolation inequality that generalizes (69) is

$$\left|\left|\left|g^{\frac{a+b}{2}}\phi\right|\right|\right|^2_{\epsilon\frac{a+b}{2}} \le c_1 \left|\left|\left|g^a\phi\right|\right|\right|_{a\epsilon} \left|\left|\left|g^b\phi\right|\right|\right|_{b\epsilon} + c\left|\left|\phi\right|\right|^2 \quad (76)$$

as long as $\epsilon\,(a+b) \le 1$. ∎

PROPOSITION 2
Let x be a boundary point of Ω. Suppose that r is a local defining function for $b\Omega$ near x. Then the (germ of the) function r is a subelliptic multiplier at x. We have the estimate

$$\left|\left|\left|r\phi\right|\right|\right|_1 \lesssim \left|\left|r\phi\right|\right|_1 \le cQ\,(\phi,\phi). \quad (77)$$

PROOF   The first inequality is trivial, because the 1-norm involves all first derivatives, and hence includes the tangential first derivatives. On the other hand, if g is any smooth function vanishing on the boundary, then two integrations by parts show that

$$\left|\left|\frac{\partial g}{\partial z_i}\right|\right|^2 = \left(\frac{\partial g}{\partial z_i},\frac{\partial g}{\partial z_i}\right) = \left(\frac{\partial g}{\partial \bar{z}_i},\frac{\partial g}{\partial \bar{z}_i}\right) = \left|\left|\frac{\partial g}{\partial \bar{z}_i}\right|\right|^2. \quad (78)$$

It suffices therefore to estimate the barred derivatives of $r\phi$. On the other hand,

$$\left|\left|r\phi\right|\right|^2_{\bar{z}} \le c\left|\left|\phi\right|\right|^2_{\bar{z}} + c\left|\left|\phi\right|\right|^2 \quad (79)$$

and these terms are included in $Q(\phi, \phi)$, so the result follows. We have used the boundedness of r on the support of ϕ, and we have controlled terms where ϕ is undifferentiated by the term $c\,\|\phi\|^2$. ∎

All these considerations would be uninteresting if one had no way of obtaining geometric information on the ideal of subelliptic multipliers. Proposition 2 reveals that the ideal of subelliptic multipliers is always nontrivial. Given one function in it, one can construct others via the procedure invented by Kohn.

Suppose that f is a subelliptic multiplier at x, and $\phi = \sum_{j=1}^{n} \phi_j d\bar{z}^j$ is the (0,1) form we wish to estimate. It turns out (Proposition 5) that

$$\left\|\left\|\sum_{j=1}^{n} \frac{\partial f}{\partial z_j} \phi_j\right\|\right\|_{\epsilon}^{2} \leq c\left(\||f\phi|\|_{2\epsilon}^{2} + Q(\phi, \phi)\right). \tag{80}$$

This shows that we can also estimate the particular function on the left of (80). In particular, if all the coordinate functions were multipliers, then the constant function unity would be also. Consider a vector field of type (1,0)

$$v = \sum_{j=1}^{n} \overline{v_j} \frac{\partial}{\partial z^j}\,. \tag{81}$$

The expression $\sum_{j=1}^{n} v_j \phi_j$ is the interior product of $\bar{v}$ with ϕ. (Recall that interior multiplication is the adjoint of exterior multiplication, so the interior multiple of a 1-form by a vector field is a function.) The formula (80) shows that if f is a real-valued subelliptic multiplier, then we can estimate the function given by interior multiplication by $\overline{\partial f}$ with ϕ. The complex conjugates are a nuisance wherever one puts them. Rather than using the terminology of interior multiplication, we introduce the concept of "allowable" vector field or "allowable" (1,0) form.

DEFINITION 3 *A vector field*

$$v = \sum_{j=1}^{n} v_j \frac{\partial}{\partial z^j}$$

of type $(1,0)$ *(or a differential form of type (1,0)* $\sum_{k=1}^{n} v_k dz^k$*) is allowable (for subelliptic estimates) if there is a neighborhood U of x, and positive constants C, ϵ such that*

$$\left\|\left\|\sum_{j=1}^{n} v_j \phi_j\right\|\right\|_{\epsilon}^{2} \leq C\, Q(\phi, \phi) \tag{82}$$

for all smooth forms $\phi = \sum \phi_j d\bar{z}^j$ *supported in U and in the domain of $\overline{\partial}^*$.*

Earlier we stated that ∂f is allowable whenever f is a subelliptic multiplier. See Proposition 5. Given any square matrix whose rows are allowable vectors, the determinant of the matrix is itself a subelliptic multiplier. See Proposition 3. Thus we have also the notion of "allowable matrix." These propositions enable us to construct more multipliers.

Let us be more precise. Write $A^{0,1} = A^{0,1}(\mathbb{C}^n)$ for the bundle of $(0,1)$ forms; we assume that these are defined near a point $x_o \in b\Omega$. Let $\mathbf{L}(A^{0,1})$ denote the bundle of linear transformations from this space to itself. A smooth local section of $\mathbf{L}(A^{0,1})$ is then a smooth mapping $x \to L_x$ from a neighborhood of x_o to the space of linear transformations at each x. If L is such a smooth section, then it makes sense to ask whether we can estimate the form $L\phi$ in terms of $Q(\phi, \phi)$.

DEFINITION 4   *A smooth local section L of $\mathbf{L}(A^{0,1})$ is called an allowable matrix at x_o if there is a neighborhood U of x_o, and positive constants C, ϵ such that*

$$|||L\phi|||_\epsilon^2 \leq C \, Q(\phi, \phi)$$

$$\phi \in \mathbf{D}_{0,1}(U). \tag{83}$$

Note that this is the same thing as saying that each row of the matrix of the linear transformation is an allowable row.

Three additional results of Kohn are as follows, although he avoids the terms allowable matrix and allowable row.

PROPOSITION 3
The determinant of an allowable matrix is a subelliptic multiplier.

PROOF   Let L be the given allowable matrix. Then $A = L^*L$ is positive semidefinite. Here L^* is the usual adjoint of a matrix on complex Euclidean space. The key step is simple linear algebra. Namely, given a positive semidefinite matrix of functions A, we can find a constant C such that

$$\det(A)\|v\|^2 \leq C(Av, v). \tag{84}$$

In other words, there is a constant C such that

$$C \, A - \det(A) \, I \tag{85}$$

is also positive semidefinite at each point. Here the norm and inner product are the usual ones on complex Euclidean space and do not involve integration. Applying this formula to $A = L^*L$, noting that $\det(A) = |\det(L)|^2$, and because

L is allowable, the result follows. Here are the details:

$$
\begin{aligned}
|||\det{(L)}\,\phi|||_\epsilon^2 &= \sum_{j=1}^{n} \int \left|\Lambda^\epsilon \det{(L)}\,\phi_j\right|^2 \\
&= \sum \int \left|\det{(L)}\,\Lambda^\epsilon \phi_j + P^{\epsilon-1}\phi_j\right|^2 \\
&\le 2\sum \int \left|P^{\epsilon-1}\phi_j\right|^2 + 2\sum \int \left|\det{(L)}\right|^2 \left|\Lambda^\epsilon \phi_j\right|^2 \\
&\le cQ\left(\phi,\phi\right) + c \int \left(L^* L\left(\Lambda^\epsilon \phi\right), \Lambda^\epsilon \phi\right) \\
&\le cQ\left(\phi,\phi\right) + c \int \left|L(\Lambda^\epsilon \phi)\right|^2 \\
&\le cQ\left(\phi,\phi\right) + c \int \left|\Lambda^\epsilon L\phi + P^{\epsilon-1}\right|^2 \\
&\le cQ\left(\phi,\phi\right) + c\,|||\phi|||_{\epsilon-1}^2 + |||L\phi|||_\epsilon^2 .
\end{aligned}
\tag{86}
$$

The last line implies that $|||\det(L)\phi|||_\epsilon^2 \le cQ(\phi,\phi)$ as long as L is allowable and ϵ is at most unity. ∎

PROPOSITION 4

The Levi form is an allowable matrix and

$$
|||\lambda\phi|||_{\frac{1}{2}}^2 \le C\,Q\left(\phi,\phi\right)
$$

$$
\phi \in \mathrm{D}_{0,1}\left(U\right).
\tag{87}
$$

PROOF To abbreviate things somewhat, we will say that a term is ok if it can be estimated by some constant times $Q\left(\phi,\phi\right)$. First we start with the definition of the tangential $1/2$ norm and the tangential elliptic pseudodifferential operator Λ. This gives the following:

$$
|||\lambda\phi|||_{\frac{1}{2}}^2 = (\lambda\phi, \Lambda\lambda\phi) = \left(\lambda\phi, \left(\Lambda^{-1} - \sum_{j=1}^{2n-1} \frac{\partial^2}{\partial t_j^2}\Lambda^{-1}\right)\lambda\phi\right).
\tag{88}
$$

The term $\left(\lambda\phi, \Lambda^{-1}\lambda\phi\right)$ is obviously ok; it remains to estimate the sum. To do so it is enough to estimate $\left(\lambda\phi, D^1 P^0 \lambda\phi\right)$ where D^1, P^0 are any first-order differential operator and zeroth-order pseudodifferential operator respectively. Suppose first that $D^1 = \overline{D}$ is a barred derivative. It follows from the Schwarz inequality and the fact that the commutator $\left[\overline{D}, P^0\lambda\right] = S^0$ is of order zero, that

$$
\left(\lambda\phi, \overline{D}P^0 \lambda\phi\right) \le c\,||\phi||\,||\phi||_{\bar z} + c\,||\phi||^2 \le c\left(||\phi||^2 + ||\phi||_{\bar z}^2\right).
\tag{89}
$$

Since both terms on the right are included in $Q(\phi, \phi)$, the whole thing is ok. If now $D^1 = D$ is a tangential unbarred derivative, then the result follows by integration by parts and similar estimates. In case the derivative is a normal derivative, one obtains also a boundary integral of the form

$$\int_{b\Omega} \left(\lambda\phi, P^0\lambda\phi\right) \, dS. \tag{90}$$

We may assume that there is a cut-off function in front of the operator of order zero. (By a cut-off function we mean a smooth nonnegative function with compact support that is equal to unity near a given set.) After one uses the Schwarz inequality on the inner product (summation) within the integral, and integrates, one sees that it remains to estimate the term

$$\int_{b\Omega} \left(\lambda P^0\phi, P^0\phi\right) \, dS. \tag{91}$$

This last term can be estimated by

$$c\,Q\left(\lambda P^0\phi, \lambda P^0\phi\right) \tag{92}$$

because of the explicit form for $Q(.,.)$. It is easy to estimate the term (92) in terms of $Q(\phi, \phi)$. For example, calling $S^0 = \lambda P^0$, we see that

$$\begin{aligned}
\left(\bar{\partial}S^0\phi, \bar{\partial}S^0\phi\right) &= \left(S^0\bar{\partial}\phi + T^0\phi, \bar{\partial}S^0\phi\right) \\
&\leq \left(c\,||\bar{\partial}\phi|| + c\,||\phi||\right) ||\bar{\partial}S^0\phi|| \tag{93}
\end{aligned}$$

so that $||\bar{\partial}S^0\phi|| \leq c\,||\bar{\partial}\phi|| + c\,||\phi||$. We have used the boundedness of operators of order zero on L^2. From this and the corresponding properties for the other terms, it follows that

$$\left|\left|\bar{\partial}\left(S^0\phi\right)\right|\right|^2 + \left|\left|\bar{\partial}^*\left(S^0\phi\right)\right|\right|^2 + ||S^0\phi||^2 \leq cQ(\phi, \phi). \tag{94}$$

This finishes the proof. ∎

PROPOSITION 5
Suppose that f is a subelliptic multiplier and

$$|||f\phi|||_{2\epsilon}^2 \leq cQ(\phi, \phi) \tag{95}$$

for all appropriate ϕ and for $0 < \epsilon \leq 1/2$. Then there is a constant $c > 0$ so that

$$\left|\left|\left|\sum_{j=1}^{n} \frac{\partial f}{\partial z_j}\phi_j\right|\right|\right|_{\epsilon}^2 \leq cQ(\phi, \phi). \tag{96}$$

In other words, ∂f is allowable.

PROOF   Let us write

$$\psi = \sum_{j=1}^{n} \frac{\partial f}{\partial z_j} \phi_j$$

so that we have

$$\left\| \sum_{j=1}^{n} \frac{\partial f}{\partial z_j} \phi_j \right\|_{\epsilon}^{2} = \sum_{j=1}^{n} \left(\Lambda^{\epsilon} \frac{\partial f}{\partial z_j} \phi_j, \Lambda^{\epsilon} \psi \right). \tag{97}$$

The idea is to rewrite everything in terms of ok terms. Doing so is somewhat complicated. We have

$$\left\| \sum_{j=1}^{n} \frac{\partial f}{\partial z_j} \phi_j \right\|_{\epsilon}^{2} = \sum_{j=1}^{n} \left(\Lambda^{\epsilon} \frac{\partial f}{\partial z_j} \phi_j, \Lambda^{\epsilon} \psi \right)$$

$$= \sum_{j=1}^{n} \left(\frac{\partial f}{\partial z_j} \phi_j, \Lambda^{2\epsilon} \psi \right)$$

$$= \sum_{j=1}^{n} \left(\frac{\partial \left(f \phi_j \right)}{\partial z_j} - f \frac{\partial \left(\phi_j \right)}{\partial z_j}, \Lambda^{2\epsilon} \psi \right)$$

$$= \sum_{j=1}^{n} \left(\frac{\partial \left(f \phi_j \right)}{\partial z_j}, \Lambda^{2\epsilon} \psi \right) + \left(f \overline{\partial}^{*} \phi, \Lambda^{2\epsilon} \psi \right)$$

$$= \sum_{j=1}^{n} \left(\frac{\partial \left(f \phi_j \right)}{\partial z_j}, \Lambda^{2\epsilon} \psi \right) + \left(\overline{\partial}^{*} \phi, \overline{f} \Lambda^{2\epsilon} \psi \right). \tag{98}$$

We have used the self-adjointness of Λ and also the formula for the adjoint of $\overline{\partial}$. Consider the last term in (98):

$$\left| \left(\overline{\partial}^{*} \phi, \overline{f} \Lambda^{2\epsilon} \psi \right) \right| \leq \left\| \overline{\partial}^{*} \phi \right\| \, \left\| \overline{f} \Lambda^{2\epsilon} \psi \right\|$$

$$= \left\| \overline{\partial}^{*} \phi \right\| \, \left\| f \Lambda^{2\epsilon} \psi \right\|$$

$$= \left\| \overline{\partial}^{*} \phi \right\| \, \left\| \Lambda^{2\epsilon} \left(f \psi \right) + P^{2\epsilon - 1} \psi \right\|$$

$$\leq \left\| \overline{\partial}^{*} \phi \right\| \left(c \left\| \| f \phi \| \right\|_{2\epsilon} + c \left\| \| \phi \| \right\|_{2\epsilon - 1} \right)$$

$$\leq c \left\| \overline{\partial}^{*} \phi \right\|^{2} + c \left\| \| f \phi \| \right\|_{2\epsilon}^{2} + c \left\| \phi \right\|^{2}. \tag{99}$$

The third line follows because the commutator is of order $2\epsilon - 1$; we are assuming that this is at most zero. The fourth line requires knowing that $\psi = \sum_{k=1}^{n} a_k \phi_k = P^0 \phi$. Finally the last line together with the hypothesis

shows that all these terms are ok. It remains to handle the remaining term from
(98). We integrate by parts; the boundary integral in (100) vanishes because ϕ
is in the domain of the adjoint, and hence

$$\sum_{j=1}^{n} \frac{\partial r}{\partial z_j} \phi_j = 0$$

on the boundary:

$$\sum_{j=1}^{n} \left(\frac{\partial (f\phi_j)}{\partial z_j}, \Lambda^{2\epsilon}\psi \right) = \sum_{j=1}^{n} \left(-f\phi_j, \frac{\partial}{\partial \bar{z}_j} \left(\Lambda^{2\epsilon}\psi \right) \right) + \sum_{j=1}^{n} \int_{b\Omega} f \frac{\partial r}{\partial z_j} \phi_j \overline{\Lambda^{2\epsilon}\psi}$$

$$= \sum_{j=1}^{n} \left(-f\phi_j, \frac{\partial}{\partial \bar{z}_j} \left(\Lambda^{2\epsilon}\psi \right) \right). \tag{100}$$

We now move the derivative past the Λ, making an error, and then throw the Λ
back over to the left. This leaves us with

$$\sum_{j=1}^{n} \left(-\Lambda^{2\epsilon} f\phi_j, \frac{\partial}{\partial \bar{z}_j} (\psi) \right) + \left(-f\phi_j, P^{2\epsilon}\phi \right)$$

$$= \sum_{j=1}^{n} \left(-\Lambda^{2\epsilon} f\phi_j, \frac{\partial}{\partial \bar{z}_j} (\psi) \right) + \left(-\Lambda^{2\epsilon} f\phi_j, P^0\phi \right)$$

$$= \sum_{j=1}^{n} \left(-\Lambda^{2\epsilon} f\phi_j, \frac{\partial}{\partial \bar{z}_j} \left(\sum_{k=1}^{n} a_k\phi_k \right) \right) + \left(-\Lambda^{2\epsilon} f\phi_j, P^0\phi \right). \tag{101}$$

We have used again the definition of ψ. Note also that $\partial/\partial \bar{z}_j$ is not tangential.
This is a minor nuisance; its normal component commutes with Λ and hence the
error $P^{2\epsilon}$ is tangential. Using the Schwarz inequality and recalling that barred
derivatives of ϕ are ok, we can estimate this by

$$c \, |||f\phi|||_{2\epsilon}^2 + c \, ||\phi||_{\bar{z}}^2 + c \, ||\phi||^2 \tag{102}$$

and hence by ok terms. This finishes the proof. ∎

6.4.3 The geometry of Kohn's algorithm

Combining the five propositions of Section 4.2 yields an algorithm for finding
subelliptic multipliers. We first describe this process in words, then we give the
precise definition. According to Proposition 4, the Levi form is an allowable
matrix. According to Proposition 3, its determinant is a subelliptic multiplier.
According to Proposition 2, the defining function r is a subelliptic multiplier.

According to Proposition 1, the ideal of subelliptic multipliers is a real radical ideal. Thus the ideal J_0, defined to be the real radical of the ideal generated by r and $\det(\lambda)$, consists of subelliptic multipliers.

If $f \in J_o$, then Proposition 5 guarantees that ∂f is an allowable row. We denote by μ_1 the collection of those allowable matrices whose rows are either rows of the Levi form or are ∂f for some $f \in J_o$. The determinants of these matrices are themselves subelliptic multipliers. We define J_1 be the real radical of the ideal generated by elements of J_0 and all determinants of elements of μ_1. When the containment $J_1 \supset J_0$ is strict, we enlarge our collection of known subelliptic multipliers. We then iterate the process by making the following definition.

DEFINITION 5 *The Kohn ideals of subelliptic multipliers are inductively defined by*

$$J_0 = \mathrm{rad}_R\left(r, \det(\lambda)\right)$$

$$J_{k+1} = \mathrm{rad}_R\left(J_k, \det\left(\mu_{k+1}\right)\right). \tag{103}$$

In (103) μ_{k+1} denotes the set of all matrices whose rows are either of the form ∂f for some $f \in J_k$ or are rows of the Levi form. Also $\det\left(\mu_{k+1}\right)$ means the ideal generated by all determinants of elements of μ_{k+1}.

Definition 5 yields a sequence of real radical ideals $J_0 \subset J_1 \subset \cdots \subset J_k \subset \cdots$ all of whose elements are subelliptic multipliers. If there is a k for which J_k is the whole ring, then the constant function unity is a subelliptic multiplier, and there is a subelliptic estimate. This is Kohn's sufficient condition for subellipticity of the $\bar{\partial}$-Neumann problem.

REMARK 3 It is possible to work solely with the allowable matrices. The analogue of a subelliptic multiplier becomes an invertible allowable matrix. The ideas remain essentially the same. ∎

Kohn discovered the geometric meaning of the ideals J_k, when the defining function is real analytic. Except where we specifically mention otherwise, in the remainder of this chapter we assume that we are in the real analytic case. We suppose that r is real analytic near the point p_0 in question. Then we may restrict ourselves to real analytic subelliptic multipliers. Kohn shows that the condition $1 \in J_k$ (i.e., J_k is the ring of germs of real analytic functions) for some k is equivalent to the nonexistence of real analytic varieties of "positive holomorphic dimension" in the boundary. We discuss such varieties in Section 5, where we give their definition and prove the following theorems.

THEOREM 8

Suppose that Ω is a smoothly bounded pseudoconvex domain in $\mathbb{C}^n$, and that its boundary is real analytic near the boundary point p_0. Let J_k denote the sequence of Kohn ideals of subelliptic multipliers at p_0. Then there is an integer k such that $1 \in J_k$ if and only if there are no varieties of positive holomorphic dimension in any neighborhood of p_0. As a consequence of this and Theorem 9, there is a subelliptic estimate for the $\bar{\partial}$-Neumann problem at p_0 if there are no complex analytic varieties of positive dimension in any neighborhood of p_0.

We prove this theorem in Section 5.2. Any complex analytic variety of positive dimension lying in the boundary is a variety of positive holomorphic dimension. Diederich and Fornaess clarified the situation by proving the following converse assertion.

THEOREM 9

Suppose that Ω is a smoothly bounded pseudoconvex domain and its boundary is real analytic near the boundary point x. Let U be a neighborhood of x. If there exists a real analytic variety W of "positive holomorphic dimension" such that $W \subset U \cap b\Omega$, then there is a (positive-dimensional) complex analytic variety V such that $V \subset U \cap b\Omega$.

We prove Theorem 9 in Section 5.3. The proofs of these theorems indicate how the concept of holomorphic dimension of real analytic varieties arises from Kohn's procedure.

Putting together Theorems 8 and 9, we see in the real analytic case that there is a subelliptic estimate at x if, for every neighborhood of x, there is no complex analytic variety lying in that neighborhood that is contained in the boundary. This is actually the same as saying that there is no (positive-dimensional) complex analytic variety containing x and lying in $b\Omega$. To see this, recall from Theorems 4.2 and 4.4 that the set of points in a real analytic hypersurface for which there is a nontrivial complex analytic variety containing the point and lying in the hypersurface is a closed set. By Theorem 4.4, in the real analytic case, such points are precisely those in the complement of the set of points of finite type. Therefore there is a subelliptic estimate near any point p of finite type on the boundary of a smoothly bounded pseudoconvex domain, if the boundary hypersurface is real analytic near p. If a bounded domain has real analytic boundary, then every boundary point is a point of finite type by Theorem 4.5.

As stated in Theorem 6, Catlin has extended the result on subellipticity to the smooth case, without using the method of subelliptic multipliers. Catlin's approach is roughly as follows. First, he uses the $\bar{\partial}$-Neumann problem with weights, as in the book [Hm]. Thus, instead of using Lebesgue measure in the integrals, one defines the norm as in (6). The weight function ψ is smooth up to the boundary and bounded above on the closed domain by unity. Given a small positive number δ, one works in the strip defined by $-\delta < r(z) \le 0$. To prove that a subelliptic estimate of order ϵ holds, one must find such a (plurisubharmonic) weight function whose Hessian is larger than $c\delta^{-2\epsilon}$ in this strip. (This means that the minimum eigenvalue is larger than $c\delta^{-2\epsilon}$.) He relates the existence of such functions to the geometry of the boundary by defining a multi-type for each boundary point. This is an n-tuple of integers or plus infinity; in the lexicographical ordering the multi-type is an upper semicontinuous function on the boundary. To prove the needed properties of the multi-type, Catlin gives a refined version of the results of Diederich-Fornaess on varieties of positive holomorphic dimension and extends the results from Chapter 4 by maintaining precise information on the size of neighborhoods and constants involved. He bounds the maximum integer in the multi-type at p by $\Delta^1(b\Omega, p)$, and uses the theorem on openness to ensure that the multi-type stays finite nearby. As stated above, the key analytic idea is the construction of plurisubharmonic functions, smooth on the closed domain and bounded by unity there, but with arbitrarily large Hessians at the boundary. Catlin observed earlier that the existence of such plurisubharmonic functions implies global regularity of the Neumann operator. To obtain the stronger statement that subellipticity holds, he must also show that the (minimum eigenvalue of the) Hessians satisfy $\partial\bar{\partial}\psi \ge c\,|r|^{-2\epsilon}$ as one approaches the boundary. Sibony [Si2] has investigated also the existence and properties of such plurisubharmonic functions. We return briefly to these considerations at the end of this chapter.

6.4.4   Some examples of Kohn's algorithm

Although we are dealing with inequalities, Kohn's process is a formal, algebraic algorithm. For simplicity we consider how the process works in the special case when the defining function has the form

$$r(z, \bar{z}) = 2\,\mathrm{Re}\,(z_{n+1}) + \sum_{j=1}^{K} |f_j(z)|^2. \tag{104}$$

For convenience, we change notation so that we are in $\mathbb{C}^{n+1}$. Suppose that the point in question is the origin and that the functions f_j are holomorphic and vanish there. For the origin to be a point of finite type (see Chapter 4), it is necessary and sufficient that the variety $V = \mathbf{V}(f_1, \ldots, f_K, z_{n+1})$ consist of the origin alone. If (104) holds, it then follows from Theorem 9 and Theorem 3.4 that there is a subelliptic estimate if and only if the (germ of the) variety V consists of the origin alone. The method yields therefore yet another equivalent condition for deciding whether a complex analytic variety is trivial.

We now amplify these comments. To simplify our computations, we assume that r is as in (104), where the f_j are independent of z_{n+1}. The resulting domain is pseudoconvex, but unbounded. We must work with a bounded domain, such that the origin is a boundary point and such that there is a neighborhood of the origin on which (104) is a local defining function. Since subellipticity is a local property, this causes no difficulty.

We begin with the Levi form. If (104) holds, the Levi form is given by $\lambda = (\partial f)^* (\partial f)$, or in coordinates, by

$$\lambda_{ij} = \sum_{m=1}^{K} \frac{\partial f_m}{\partial z_i} \overline{\frac{\partial f_m}{\partial z_j}} \; . \tag{105}$$

It follows that its determinant satisfies

$$\det(\lambda) = \sum \left| J\left(f_{i_1}, \ldots f_{i_n}\right) \right|^2 \tag{106}$$

where the sum is taken over all possible choices of n of the functions, and J denotes the Jacobian determinant.

The first ideal in the Kohn process is the radical of the ideal generated by the defining function and the determinant of the Levi form. Because of Proposition 1, the first ideal contains each of the Jacobian determinants in (106). From these considerations we see that the process is equivalent to a simpler one that involves only holomorphic functions. We make the simpler process into a definition. Recall that the holomorphic radical is defined in Chapter 2.

DEFINITION 6   *Let (f) be an ideal in $_nO_o$. We associate with (f) a nested sequence of radical ideals $I_1 \subset I_2 \ldots$ in $_nO_o$. For each choice of $g_1, \ldots, g_n \varepsilon (f)$ consider the derivative matrix of the corresponding mapping:*

$$\begin{pmatrix} \frac{\partial g_1}{\partial z_1} & \cdots & \frac{\partial g_1}{\partial z_n} \\ \cdots & \cdots & \cdots \\ \frac{\partial g_n}{\partial z_1} & \cdots & \frac{\partial g_n}{\partial z_n} \end{pmatrix} . \tag{107}$$

We compute determinants $J(g)$ of all such matrices. These determinants generate a new ideal, say $I_1^\#$. We define

$$I_1 = \mathrm{rad}\left(I_1^\#\right) . \tag{108}$$

Consider the ideal $I_2^\#$ generated by I_1 together with all determinants of square matrices of first derivatives of elements in either (f) or I_1. By taking its radical we define an ideal $I_2 = \mathrm{rad}(I_2^\#)$. We continue this process and define thereby a nested sequence of radical ideals in O. Notice that the original ideal is not a member of the nested sequence.

It is important to emphasize that we do not include the original generators in the ideals I_k. Doing so would not conform to the Kohn algorithm; furthermore, the process would end too easily. To see this, suppose that the variety of the original ideal is trivial. Then the radical of the original ideal would be the maximal ideal; the next step would yield the constant function unity as a Jacobian determinant. Thus, we are allowed to use the derivatives of the original generators as allowable rows, but we are not allowed to use the generators themselves as multipliers.

Our comments so far imply that the original ideal (f) defines a zero-dimensional variety if and only if this process yields the full ring in finitely many steps; that is, there is an integer k such that $1 \in I_k$. We illustrate the process in the next two examples.

Example 3

Suppose $n = 2$, our variables are (z, w) and our ideal is

$$\left(z^2 - w^3, zw\right). \tag{109}$$

The Jacobian is $2z^2 + 3w^3$. Its derivatives, together with the derivatives of the generators, give us these allowable rows:

$$\begin{pmatrix} 2z & -3w^2 \\ w & z \\ 4z & 9w^2 \end{pmatrix}. \tag{110}$$

From (110) we obtain the determinant function $9w^3 - 4z^2$. Subtracting appropriate multiples of this from the original determinant gives us the functions z^2 and w^3. Now, taking radicals, we obtain the coordinate functions. One more determinant (of the Jacobian of the identity transformation (z, w)) gives us the constant unity. □

Example 4

Let us now suppose that the ideal is generated by a regular sequence of order n of the following form:

$$f_1(z) = z_1^{m_1}$$
$$f_2(z) = z_2^{m_2} + a_1(z_1) z_2^{m_2-1} + \cdots + a_{m_2}(z_1)$$

$$\cdots\cdots$$

$$f_n(z) = z_n^{m_n} + \cdots + b_{m_n}(z_1, \ldots, z_{n-1}). \tag{111}$$

To be more precise, we suppose that each f_i is a Weierstrass polynomial of degree m_i in z_i with coefficients in $(z_1, \ldots, z_{i-1})$. Given any nonzero ideal, one can make a linear change of coordinates so that there exist functions in the ideal with these properties. This we did explicitly in Section 1.4.2. If the dimension of the variety of the ideal is zero, then one can continue this process until the number of functions is n. Here we make the additional assumption that these functions generate the ideal. Let $B = df$ be the matrix (112) below. We comment that, if the defining function were $2\text{Re}(z_{n+1}) + \|f(z)\|^2$, then the Levi form would satisfy $\lambda = B^*B$. For the holomorphic version, we begin with the lower triangular derivative matrix B:

$$\begin{pmatrix} m_1 z_1^{m_1-1} & 0 & \cdots & 0 \\ & \frac{\partial f_2}{\partial z_2} & \cdots & 0 \\ & * & \cdots & 0 \\ & * & * & \frac{\partial f_n}{\partial z_n} \end{pmatrix}. \tag{112}$$

Note that the determinant contains a factor of $z_1^{m_1-1}$. Hence, when we take radicals, we see that $g = z_1 \prod_{j=2}^n (\partial f_j / \partial z_j)$ is in I_1. Only the last term in the product depends on z_n. Replacing the last row by the allowable row ∂g and taking determinants, we obtain a multiplier with a factor of $(\partial / \partial z_n)^2 f_n$. We next use its gradient as the last allowable row and continue this process. Since f_n is a Weierstrass polynomial of degree m_n, after m_n differentiations we obtain a constant. This yields the multiplier $h = z_1 \prod_{j=2}^{n-1} (\partial f_j / \partial z_j)$, which is independent of z_n. We use ∂h for the penultimate allowable row. After taking determinants and radicals, we obtain that

$$t = z_1 \prod_{j=2}^{n-2} \frac{\partial f_j}{\partial z_j} \left(\frac{\partial}{\partial z_{n-1}} \right)^2 f_{n-1} \frac{\partial f_n}{\partial z_n}$$

is a multiplier. We apply the same process to obtain eventually that

$$w = z_1 \prod_{j=2}^{n-2} \frac{\partial f_j}{\partial z_j} \left(\frac{\partial}{\partial z_{n-1}} \right)^2 f_{n-1}$$

is a multiplier. Now we use ∂w as the penultimate allowable row and repeat. Since each f_j contains a term of the form $z_j^{m_j}$, we can systematically eliminate all the other variables until we have only z_1 left. Then we discover that z_1 is a multiplier, and hence $(1, 0, \ldots, 0)$ is an allowable row. After all these steps, we may finally ignore the role of z_1 by setting it equal to zero. This gives a reduction in the dimension, enabling us to prove by induction that all the coordinate functions are multipliers. At that point the next determinant gives us the constant function unity.

The reader will benefit by following this proof in the simple case where each f_j is simply $z_j^{m_j}$; the first determinant is the product of the coordinate functions. The next step obtains each of the products of all but one of the coordinate functions as multipliers. Eventually we reach the coordinate functions themselves. $\square$

Example 4 illustrates some interesting things. Suppose that $r = 2 \operatorname{Re}(z_{n+1}) + \|f(z)\|^2 - \|g(z)\|^2$ where $z = (z_1, \ldots, z_n)$. The Levi form can then be written $B^*B - C^*C$ where B, C have holomorphic entries. Even when $C = 0$, the Levi form may not be diagonalizable. This is so in Example 4, but the matrix B

is lower triangular and easy to understand. In general, however, these matrices are neither finite-dimensional, nor lower triangular.

When (f) is nice, such as in Examples 3 and 4, one can say more. It is possible to determine how many steps the process takes and what powers are involved in taking the radicals, and give thereby an estimate on the value of epsilon for the domain defined near the origin by

$$r(z, \bar{z}) = 2 \operatorname{Re} (z_{n+1}) + \sum_{j=1}^{K} |f_j(z)|^2. \tag{113}$$

The author believes that interesting mathematics would result if someone were to analyze in full depth the relationships among the process in Definition 6, Theorem 1, and formulas (82) and (84) from Chapter 2. Roughly speaking, in each context, the Jacobian determinant is that function not in the ideal generated by the given functions, but closest to being in the ideal. More precisely, suppose that we are given a regular sequence $(f_1, \ldots, f_n)$ in O_n of order n. Then

$$\det \left(\frac{\partial f_i}{\partial z_j} \right) \not\equiv 0 \mod (f) \tag{114}$$

but

$$g \det \left(\frac{\partial f_i}{\partial z_j} \right) \equiv 0 \mod (f) \tag{115}$$

whenever g vanishes at the origin. This shows that the determinant generates the "socle" of the ideal. See [EL] for more on the socle. On the other hand, given a domain defined by (104), the determinant of the Levi form is the subelliptic multiplier that enables us to begin. The determinant equals

$$\left| \det \left(\frac{\partial f_i}{\partial z_j} \right) \right|^2. \tag{116}$$

In some sense it is the function that vanishes on the largest possible set yet still works as the first step in a process that leads to a nonvanishing function. It seems interesting also to extend these considerations to formal power series.

6.5 Varieties of positive holomorphic dimension

6.5.1 Introduction

We return to the geometry in the real analytic case. Fix a boundary point p_0. Kohn's process yields an increasing sequence of real radical ideals

$$J_0 \subset J_1 \subset J_2 \subset \cdots \tag{117}$$

of subelliptic multipliers. These are ideals in the ring of germs at p_0 of real analytic functions. Corresponding to (117) is a decreasing sequence of real analytic varieties

$$\mathbf{V}(J_0) \supset \mathbf{V}(J_1) \supset \mathbf{V}(J_2) \supset \cdots . \tag{118}$$

If $\mathbf{V}(J_k)$ is empty for some k, then we have a subelliptic estimate at p_0.

The first variety $\mathbf{V}(J_0)$ consists of the points where the Levi form degenerates; we suppose that p_0 is such a point. If it were isolated, then the coordinate functions would be in J_0, since J_0 is a radical ideal. Taking the differentials ∂z^j of the coordinate functions as allowable rows shows that the identity matrix is allowable; thus $\mathbf{V}(J_1)$ is empty, and we obtain a subelliptic estimate. Thus a subelliptic estimate holds at points of isolated degeneracy of the Levi form, in the real analytic case. The same conclusion is not possible in the smooth case.

More generally, suppose that $\mathbf{V}(J_0)$ is positive dimensional. If every nonzero $(1,0)$ vector field L, tangent to $\mathbf{V}(J_0)$, were also in the kernel of the Levi form, then the process breaks down. More precisely, for any $f \in J_0$, $L(f) = \langle df, L \rangle = \langle \partial f, L \rangle = 0$. Replacing a row of the Levi form by ∂f does not help, as L remains in the kernel of the new matrix. Any determinant of such an allowable matrix must vanish on $\mathbf{V}(J_0)$. Thus

$$J_1 \subset \mathbf{I}(\mathbf{V}(J_0)) = \text{rad}_R(J_0) = J_0 \subset J_1 \tag{119}$$

and we get nowhere.

This leads us to the notion of varieties of positive holomorphic dimension. We cannot have certain positive dimensional real analytic varieties W sitting in the boundary. They are called varieties of "positive holomorphic dimension"; see Definition 8 for the precise notion. A variety W has "positive holomorphic dimension" if there exists a $(1,0)$ vector field L tangent to W that lies also in the kernel of the Levi form along W. In the case of pseudoconvex boundaries, these varieties are closely related to, but need not be, complex analytic varieties of positive dimension. We prove the theorem of Diederich and Fornaess to this effect in Section 5.3.

We begin our development of holomorphic dimension. Suppose that V is a real analytic real subvariety of $\mathbb{C}^n$, containing the point p. We can extend many of the differential geometric properties from Chapter 3 to this situation; in particular we can talk about its tangent spaces. Because we will be considering singularities, we also define the "tangent space" of an arbitrary ideal.

DEFINITION 7 *Suppose that p_0 is a point and I_{p_0} is an ideal of germs of real analytic functions there. For points p on the variety $\mathbf{V}(I_{p_0})$, the Zariski tangent space of the ideal is defined by*

$$T_p^{1,0}(I_{p_0}) = \left\{ L = \sum_{j=1}^{n} a_j \frac{\partial}{\partial z^j} : L(f)(p) = 0 \quad \forall f \in I_{p_0} \right\}. \tag{120}$$

The Zariski tangent space of a real analytic real subvariety V of $\mathbb{C}^n$, at the point p, is the complex vector space $T_p^{1,0}(V) = T_p^{1,0}(\mathbf{I_p}(V))$. Here $\mathbf{I_p}(V)$ is the ideal of germs at p of real analytic functions vanishing on V.

Observe that, for manifold points, this is the same definition of $(1,0)$ tangent vector given in Chapter 3. At singular points, the space is higher dimensional; it is possible, for example, that every $(1,0)$ vector may be an element of the Zariski tangent space.

The real analytic varieties that we consider will lie in the (real analytic) boundary of a pseudoconvex domain. Suppose therefore that V is a real analytic subvariety lying within a real analytic hypersurface M. How do we decide whether V is a complex analytic variety? For the application to subelliptic estimates, the real hypersurface is pseudoconvex. The present methods rely on the semidefiniteness of the Levi form, so we hypothesize henceforth that M is pseudoconvex.

To motivate the discussion, suppose that V is a complex analytic variety of dimension q, lying in M, and that p is a nonsingular point. We saw in Chapter 3 that there are q linearly independent vectors in the kernel of the Levi form there. Let us augment that discussion. We work near a point p where V is a manifold. We may suppose that local coordinates are chosen so that p is the origin and $V = \{z : z^{q+1} = \cdots = z^n = 0\}$. Let r be a local defining function for M. We may assume that r is chosen so that

$$r(z, \bar{z}) = 2 \operatorname{Re}(z^n) + \cdots \tag{121}$$

where the dots denote higher order terms that include no pure terms. According to Theorem 4.3, we know that z^n must vanish on the variety. As in Chapter 3, let

$$L_j = \frac{\partial}{\partial z^j} - \frac{r_j}{r_n} \frac{\partial}{\partial z^n}$$

be a local basis for $T^{1,0}M$. These vector fields lie in the Zariski tangent space at each point of the variety. To see that $L_j \in T_p^{1,0}V$, $j = 1, \ldots, q$, we must verify that

$$L_j(z^k) = 0, \qquad q+1 \leq k \leq n. \tag{122}$$

This is obvious except when $k = n$. In this case, we see that

$$L_j(z^n) = -\frac{r_{z_j}}{r_{z_n}}. \tag{123}$$

The expression r_{z_j} does not vanish identically, but it does vanish along the variety V. To see this, write the local coordinates $z = (z^1, \ldots, z^q, z^{q+1}, \ldots z^n) = (t, s)$ where t stands for the first q of the coordinates. Then the function

$$t = (t^1, \ldots, t^q) \to r(t, 0, \bar{t}, 0) \tag{124}$$

vanishes identically, since $V \subset M$. Differentiating this function with respect to t_j and using the chain rule implies what we want.

We claim further that the Levi form (on M) applied to each of these vector fields vanishes at p, and each of them is of infinite type in the sense of commutators. To see this we use Proposition 3.1. From that formula for the Levi form, where subscripts denote partial derivatives, we see that

$$\lambda\left(L_j, \overline{L_j}\right) = \frac{r_{j\bar{j}}\,|r_n|^2 - 2\,\mathrm{Re}\left(r_{j\bar{n}}r_n r_{\bar{j}}\right) + r_{n\bar{n}}\,|r_j|^2}{|r_n|^2}. \tag{125}$$

By the above differentiation r_j vanishes along the variety. Differentiating again with respect to the conjugated variable $\bar{t}_j$ shows that $r_{j\bar{j}}$ also vanishes there. By (125), so does the Levi form. Furthermore, the method of computation given in the proof of Theorem 4.9 implies also that the vector field is of infinite type, as the contraction with η of any iterated commutator involves only tangential derivatives of functions vanishing on V.

As in Chapter 3, we record the basic point. If there is a q-dimensional complex manifold in a real hypersurface, then we can find q linearly independent $(1,0)$ vector fields that are tangent to the manifold and in the kernel of the Levi form along the manifold. This and the remarks on subelliptic multipliers in the first two paragraphs of this section lead us to Definition 8 below, where the complication arises from the necessity to consider singularities.

We denote the Levi form $\lambda_p\left(L, \overline{L}\right)$ by $\lambda_p(L)$ for notational ease, and we write N_p for its kernel at p. The hypothesis of pseudoconvexity guarantees that the kernel of the Levi form is a linear space. If a matrix A is not positive semidefinite, then the equations $0 = (Av, v) = (Aw, w)$ do not imply that $0 = (A(v+w), v+w)$. If A is positive semidefinite, then the implication does hold. One easy way to check this is to consider the square root of A.

We follow Kohn in giving the definition of holomorphic dimension.

DEFINITION 8 *The holomorphic dimension of a real analytic subvariety V contained in a pseudoconvex real hypersurface M is the minimum over the variety of the pointwise holomorphic dimension:*

$$\mathrm{Dim}^O(V) = \min_p\left(\mathrm{Dim}_p^O(V)\right). \tag{126}$$

The pointwise holomorphic dimension is defined by

$$\mathrm{Dim}_p^O(V) = \sup\inf\left(\dim\left(T_x^{1,0}(V) \cap N_x\right)\right). \tag{127}$$

Here the supremum is taken over all neighborhoods of p, the infimum is taken over all points in such a neighborhood, and the dimension is that of complex vector spaces.

We have verified above that a complex manifold of dimension q (and lying in M) has holomorphic dimension q at each of its points also. To say that the holomorphic dimension of a variety at a point is q implies that there is a nearby regular point where the holomorphic dimension is also q. Furthermore, one can choose the regular point x_o so that

$$\dim \left(T_x^{1,0}(V) \cap N_x \right) = q \tag{128}$$

for every point x on the variety and sufficiently close to x_o.

We emphasize what Definition 8 says when q equals unity. A variety has "positive holomorphic dimension" at a point x_o if, for every sufficiently close point x of the variety, there is a nonzero vector

$$L = \sum_{j=1}^{n} a_j \frac{\partial}{\partial z^j}$$

such that both (129) and (130) are satisfied:

$$\sum_{j=1}^{n} a_j \frac{\partial f}{\partial zj}(x) = 0 \tag{129}$$

whenever f vanishes on V and

$$\sum_{i,j=1}^{n} r_{z_i \bar{z}_j}(x) \, a_i \overline{a_j} = 0. \tag{130}$$

It is worth remarking that it is not good enough to work just at the point x_o. If, for example, the Levi form vanishes completely at this point, then every vector satisfies the second condition there.

The condition of positive holomorphic dimension for a given variety is sufficient to guarantee that there is a complex variety lying in the (pseudoconvex) hypersurface. The examples below show, however, that the given variety need not itself be complex.

Example 5

Suppose that M is given by the defining function $2 \operatorname{Re}(z^n)$ and V is given by the defining functions

$$\sum_{j=1}^{n} \left| z^j \right|^2 - 1$$
$$2 \operatorname{Re}(z^n). \tag{131}$$

Thus M is completely Levi flat and V is the intersection of the unit sphere with M. Since the Levi form on M vanishes identically, every $(1,0)$ vector field tangent to V is in the kernel of the Levi form. Since V is of real dimension $2n - 2$, and everywhere a smooth CR manifold with CR codimension equal to two, there are $n - 2$ linearly independent such vector fields. Hence V has holomorphic dimension $n - 2$ in M. We have seen many times that V contains no complex analytic varieties of positive dimension. In particular V is not a complex analytic variety. We observe, however, that through any point of V, there is a complex manifold (namely a copy of $\mathbb{C}^{n-1}$) passing through that point and lying in M. ☐

To close this introduction we indicate some consequences of the statement that there is a (1,0) vector field L such that the Levi form vanishes along some CR submanifold. Suppose that M is any pseudoconvex CR manifold, $W \subset M$ is a CR submanifold, $L \in T^{1,0}W$, and

$$\lambda\left(L, \overline{L}\right) = 0 \tag{132}$$

along W. We assume that λ represents the Levi form on M and that α is the differential 1-form defined by Lemma 3.1. We write

$$[L, \overline{L}] = B - \overline{B} + \lambda\left(L, \overline{L}\right) T. \tag{133}$$

Simple computation (apply the Jacobi identity to $\left[[L, \overline{L}], \overline{B}\right]$ and contract with η) reveals that

$$\lambda\left(B, \overline{B}\right) = \lambda\left(L, \overline{[L, B]}\right) + \lambda\left(\pi_{10}\left[\overline{B}, L\right], \overline{L}\right)$$
$$+ \overline{(\alpha_L - L)}\lambda\left(L, \overline{B}\right) + \overline{(\alpha_B - B)}\lambda\left(L, \overline{L}\right). \tag{134}$$

The Cauchy-Schwarz inequality for a positive semidefinite form gives us the result that

$$\left|\lambda\left(L, \overline{A}\right)\right|^2 \leq \lambda\left(L, \overline{L}\right) \lambda\left(A, \overline{A}\right) \tag{135}$$

for any (1,0) vector fields. Assuming that $\lambda\left(L, \overline{L}\right)$ vanishes along W, we conclude from (135) that the first three terms on the right of (134) all vanish there also. The last term is another matter. It vanishes if B is tangent to W, but not in general. We give momentarily an example where B is not tangent. If it is tangent, however, then we can derive more information. Let $A = [B, L]$ denote the commutator. It is then also tangent, and its Levi form vanishes along W also. This follows from the formula

$$\lambda\left(A, \overline{A}\right) = -\lambda\left(L, \pi_{01}\left[B, \overline{A}\right]\right) + \lambda\left(B, \pi_{01}\left[L, \overline{A}\right]\right)$$
$$+ (\alpha_L - L)\lambda\left(B, \overline{A}\right) - (\alpha_B - B)\lambda\left(L, \overline{A}\right) \tag{136}$$

the tangency of the vector fields and (135). Similar results apply to the (1,0) parts of higher order commutators. The next example shows that one must be careful.

Example 6

Suppose that M is given by $\mathrm{Re}\,(z_3) = 0$ and that W is the CR submanifold given by $\mathrm{Re}\,(z_3) = 2\,\mathrm{Re}\,(z_2) - |z_1|^2 = 0$. Then the Levi form on M vanishes identically. Let

$$L = \frac{\partial}{\partial z_1} + \overline{z_1}\frac{\partial}{\partial z_2}\,. \tag{137}$$

A calculation reveals that $L \in T^{1,0}W$, the Levi form on it obviously vanishes, and yet the vector field $B = \pi_{10}\left[L, \overline{L}\right] = -\partial/\partial z_2$ is not tangent to the submanifold W. Since W has positive holomorphic dimension, we will alter it and find a related complex manifold and $(1,0)$ tangent vector field. The new vector field will be tangent to a holomorphic curve. In this example we see that the coordinate vector field $\partial/\partial z_1$ does what we want. ☐

6.5.2   The relationship to subelliptic estimates

The concept of holomorphic dimension arose from the geometry of Kohn's ideals. We therefore turn to the proof of Theorem 8. Suppose that J is an ideal in the ring of real analytic germs at p_o. We write $\mathbf{V}\,(J)$ for the real analytic variety defined by J. If W is a germ of a real analytic variety, then we denote by $\mathbf{I}\,(W)$ the ideal of the variety. As stated in Theorem 7, we have the Lojasiewicz result that $\mathrm{rad}_{\mathbf{R}}\,(J) = \mathbf{I}\,(\mathbf{V}\,(J))$. At times we will need to consider $\mathbf{I}_{\mathbf{p}}\left(\mathbf{V}\left(J_{p_o}\right)\right)$. By this we mean that the ideal J_{p_o} is based at p_o, but we look at germs at another point p when we consider the ideal of the variety.

It is a standard difficulty in the real analytic category that $\mathbf{I}_{\mathbf{p}}\left(\mathbf{V}\left(J_{p_0}\right)\right)$ is not necessarily generated by germs at p_0, for p close to p_0 [M]. Thus the ideal sheaf is not coherent. Suppose that $J_{p_0} = \mathrm{rad}_{\mathbf{R}}\left(J_{p_0}\right)$ is a radical ideal. Although coherence fails, the following weaker statement does hold.

PROPERTY 1

There is a sequence of points converging to p_o such that the following holds. Each point in the sequence has a neighborhood on which every point p has the property that $\mathbf{I}_{\mathbf{p}}\left(\mathbf{V}\left(J_{p_0}\right)\right)$ is generated by the germs at p_o.

Property 1 is not hard to check, using the complexification of the ideals and the coherence of ideal sheaves in the complex analytic category. See [K3, p. 111] for more on this point.

The radical ideals we will use are the Kohn ideals from Definition 5. They are ideals of real analytic subelliptic multipliers on $(0,1)$ forms. Fix the point p_o in the boundary of a smoothly bounded pseudoconvex domain and suppose that the boundary is real analytic near there. The process of subelliptic multipliers yields a sequence of real radical ideals $J_0 \subset J_1 \subset \cdots \subset J_k \subset \cdots$. A subelliptic estimate holds at the point p_o if a nonvanishing function occurs after finitely many steps, that is, if there is a k for which J_k is the whole ring there.

The variety $\mathbf{V}(J_0)$ consists precisely of the weakly pseudoconvex boundary points. If this contains a manifold, then pass to a regular point p on it. If this manifold is of positive holomorphic dimension, then we can find a $(1,0)$ vector L there such that both $L \in N_p$ and also $L(f)_p = 0$ whenever f lies in J_0. Any new allowable matrix must contain an allowable row defined by ∂f for some such f. Hence any resulting determinant must vanish at p. Thus $p \in \mathbf{V}(J_1)$ also. Suppose inductively that $p \in \mathbf{V}(J_k)$. Then $p \in \mathbf{V}(J_{k+1})$ means that there is a nonzero vector L such that

$$\lambda(L, \cdot)_p = 0$$
$$L(f)_p = 0 \qquad f \in J_k. \tag{138}$$

These two conditions mean that L is both in the kernel N_p of the Levi form and it is tangent to the variety $\mathbf{V}(J_k)$. Thus, if some $\mathbf{V}(J_k)$ has positive holomorphic dimension, then the process stabilizes and defines a nonempty variety.

On the other hand, suppose there are no varieties of positive holomorphic dimension in the boundary. We claim that each iteration of the process decreases the dimension, that is,

$$\dim(\mathbf{V}(J_k)) > \dim\left(\mathbf{V}(J_{k+1})\right) \tag{139}$$

until we reach the empty variety. If the claim were false, then the dimensions would be equal, and we could pass to generic regular points where the varieties are of maximal dimension. Since there is containment, there would then be equality of the germs there, $\mathbf{V}(J_k) = \mathbf{V}(J_{k+1})$, and hence $\mathbf{I}_p(\mathbf{V}(J_k)) = \mathbf{I}_p(\mathbf{V}(J_{k+1}))$ at such points. By Property 1 on coherence, there is a sequence of points converging to p_0 so that $\mathbf{I}_{p_m}(\mathbf{V}(J_k)) = \mathbf{I}_{p_m}(\mathbf{V}(J_{k+1}))$ is generated by J_k for elements p_m of this sequence. Furthermore, there are open neighborhoods of these points on which the same holds. Recall that the ideals J_k are radical ideals by Proposition 1. On an open subset of points p arbitrarily close to p_0, we see that

$$J_{k+1} = \mathbf{I}_p\left(\mathbf{V}(J_{k+1})\right) = \mathbf{I}_p(\mathbf{V}(J_k)) = \mathrm{rad}_R(J_k) = J_k. \tag{140}$$

Thus the ideals are the same at the point p_0, so the process of adjoining new allowable rows in going from $k \to k+1$ didn't get us anywhere. This means that there is a solution L to the equations $\lambda(L, \bar{L})_p = 0$ and $L(f)_p = 0$ for all $f \in J_k$. This contradicts the hypothesis that there is no variety of positive holomorphic dimension in any neighborhood of p_0. Thus the dimensions must decrease when there are no varieties of positive holomorphic dimension in the boundary. We then obtain eventually the ideal of the empty variety, that is, the whole ring. Hence a subelliptic estimate holds.

If the process never yields an empty variety, then we have seen that it stabilizes and unearths a variety of positive holomorphic dimension that lies in the boundary. Conversely, one can also check that if a variety of positive holomorphic dimension appears in the boundary, then it must be contained in $\mathbf{V}(J_k)$ for each k. This is clear for $\mathbf{V}(J_0)$ and follows by induction. We conclude that, in the real analytic case, the process stops. There are two possibilities: In the first case $(J_{2n}) = (1)$ is the whole ring. In the second case $\mathbf{V}(J_k) = \mathbf{V}(J_{k+1})$ is a variety of positive holomorphic dimension lying in the boundary. We have proved a fundamental theorem of Kohn.

THEOREM 8
Suppose that Ω is a smoothly bounded pseudoconvex domain in $\mathbb{C}^n$, and that its boundary is real analytic near the boundary point p_0. Let J_k denote the sequence of ideals of subelliptic multipliers at p_0. Then there is an integer k such that $1 \in J_k$ if and only if there are no varieties of positive holomorphic dimension in any neighborhood of p_0. As a consequence of this and Theorem 9, there is a subelliptic estimate for the $\bar{\partial}$-Neumann problem at p_0 if there are no complex analytic varieties of positive dimension in any neighborhood of p_0.

The converse of the last statement is also true. If there is a complex analytic curve in the boundary, then it is not difficult to verify that local regularity for the canonical solution of the Cauchy–Riemann equations fails. As local regularity is a consequence of subelliptic estimates, the result follows. See [K3, p. 120] for the failure of local regularity, under the additional hypothesis of a local separating function.

By combining Theorem 8, the Diederich–Fornaess Theorem 9, and Theorem 4.4, we obtain the special case of Catlin's Theorem 6 when the boundary is real analytic near p_0. It is worth remarking that one important question remains open. It is clear that the containment $1 \in J_k$ implies subellipticity in the smooth case. Also the definition implies that subellipticity means that the function unity is a subelliptic multiplier. It is not known whether subellipticity forces this in finitely many iterations of the process. Must $1 \in J_k$ hold for some k when there is a subelliptic estimate? In other words, is the condition that there be an integer k for which $1 \in J_k$ equivalent to the notion of finite type?

In Catlin's approach to the estimates, the concept of holomorphic dimension arises also. The idea is that manifolds of holomorphic dimension zero are good. One shows that the set of weakly pseudoconvex points is contained in a union of compact sets, each contained in a manifold of holomorphic dimension zero. Submanifolds of holomorphic dimension zero are B-regular, in the language of Sibony. A compact set K in $\mathbb{C}^n$ is B-regular if, for each point $p \in K$, there is a plurisubharmonic function ϕ such that

$$\phi(p) = 0$$

$$\phi(z) < 0 \qquad z \in K - \{p\}. \tag{141}$$

See [Si1,Si2] for the precise meaning of plurisubharmonic in this context and for equivalences. One equivalence is the existence of smooth bounded plurisubharmonic functions with arbitrarily large Hessians. In case the compact set is a submanifold of holomorphic dimension zero in the boundary of a pseudoconvex domain, Catlin [C1] proved that such functions exist. This fact is required in his proof of the subelliptic estimates. Sibony's concept of B-regularity for a compact set is one manner of expressing the property that there is no complex analytic material in that set. It seems that quantitative versions of this property are fundamental in many related questions; this is perhaps the main point of this entire book.

6.5.3   The Diederich–Fornaess theorem

We complete the chapter by proving the theorem of Diederich and Fornaess that finds complex varieties in a pseudoconvex real hypersurface when there are real varieties of positive holomorphic dimension in the hypersurface. We recall its precise statement.

THEOREM 9
Suppose that M is a real analytic, pseudoconvex hypersurface in $\mathbb{C}^n$ and $V \subset M$ is a real analytic subvariety of positive holomorphic dimension q. Suppose that $p \in V$ is an arbitrary point and that U_p is a neighborhood of p in $\mathbb{C}^n$. Then we can find a complex analytic subvariety $W \subset U_p \cap M$ such that $W \cap V$ has holomorphic dimension q and, furthermore, the variety W contains p.

From Examples 5 and 6 we see that it is not possible to conclude in general that V is itself a complex analytic variety. Also, in the original statement of this theorem, the information that the variety passes through the given point was not available. This conclusion follows from combining the theorem on openness with Theorem 4.4, and because the neighborhood U_p is arbitrary. The complex analytic variety we find will be, in general, singular at the point p.

We begin with some lemmas that clarify the relationship between the real and complex geometry. Note that Lemma 3 is closely related to the remarks before Example 6. Compare it also with Lemma 2 of Chapter 3.

LEMMA 3
Suppose that M is a smooth pseudoconvex real CR manifold of hypersurface type. (The reader may assume it is a smooth pseudoconvex real hypersurface in $\mathbb{C}^n$.) Suppose further that $V \subset M$ is a real submanifold, and that X, Y are real vector fields tangent to V that have the following properties:

1.   *There are $(1,0)$ vector fields L, A on M such that $X = 2 \operatorname{Re}(L)$, $Y = 2 \operatorname{Re}(A)$.*

2.   *At points of V, the Levi forms vanish:*

$$\lambda\left(L, \overline{L}\right) = \lambda\left(A, \overline{A}\right) = 0. \tag{142}$$

Then there is a vector field B of type $(1,0)$ on M such that along V,

1. $[X, Y] = 2 \operatorname{Re}(B)$.
2. $\lambda(B, \overline{B}) = 0$.

PROOF Begin by writing

$$[X, Y] = [L + \overline{L}, A + \overline{A}] = 2 \operatorname{Re}[L, A] + 2 \operatorname{Re}[L, \overline{A}]. \tag{143}$$

By the integrability of $T^{1,0}(M)$, $[L, A]$ is a $(1,0)$ vector field. The other term can be written

$$[L, \overline{A}] = \pi_{10}[L, \overline{A}] + \pi_{01}[L, \overline{A}] + \lambda(L, \overline{A}) T. \tag{144}$$

Observe that

$$\left| \lambda(L, \overline{A}) \right|^2 \leq \lambda(L, \overline{L}) \, \lambda(A, \overline{A}) \tag{145}$$

since the Levi form is positive semidefinite. Since each term on the right of (145) vanishes along V, so does the term on the left. Putting this in (143) enables us to write the commutator as the real part of a $(1,0)$ vector field:

$$
\begin{aligned}
[X, Y] &= 2 \operatorname{Re} \left([L, A] + \pi_{10}[L, \overline{A}] + \overline{\pi_{01}[L, \overline{A}]} \right) \\
&= 2 \operatorname{Re}(B).
\end{aligned} \tag{146}
$$

This proves statement 1. Now we must verify that the Levi form vanishes. According to Definition 3.2 and also using the Jacobi identity, we can write

$$
\begin{aligned}
\lambda(B, \overline{B}) &= \langle \eta, [B, \overline{B}] \rangle \\
&= \langle \eta, [(B + \overline{B}), \overline{B}] \rangle \\
&= \langle \eta, [[X, Y], \overline{B}] \rangle \\
&= -\langle \eta, [[Y, \overline{B}], X] \rangle - \langle \eta, [[\overline{B}, X], Y] \rangle.
\end{aligned} \tag{147}
$$

We claim that each of the last two terms vanishes. To see this, replace X by $L + \overline{L}$ and Y by $A + \overline{A}$. Then each term involves a Levi form for which one of the vectors is $L, \overline{L}, A, \overline{A}$, and hence the result vanishes by (142) and (145). Recall also that the 1-form η annihilates tangent vectors of type $(1,0)$ and type $(0,1)$. ∎

The next lemma goes back to Rossi. It concerns real analytic CR submanifolds of $\mathbb{C}^n$ of arbitrary codimension. We want to know into what perhaps lower dimensional $\mathbb{C}^k$ we can locally put such manifolds.

LEMMA 4

Suppose that $V \subset \mathbb{C}^n$ is a real submanifold, $T^{1,0}(V)$ has constant rank q, and V itself has dimension $2q + k$. For each point $p \in V$, there is a neighborhood U_p and a biholomorphic mapping $\phi : U_p \to \mathbb{C}^n$ such that

1. $\phi(p) = 0$

2. $\phi(V \cap U_p) \subset \mathbb{C}^{q+k} \times \{0\}$.

PROOF   Without further comment we work with the germs at p of both manifolds and functions. After a translation and then a linear change of coordinates, we may assume that p is the origin and the tangent space there lines up properly. Hence we choose coordinates $z = (z', z'')$, where $z' = (z^1, \ldots, z^{q+k})$, so that $T_p V \approx \mathbb{C}^q \times \mathbb{R}^k$. To prove the lemma, we want to write

$$\phi(z', z'') = (z', z'' - f(z')) \tag{148}$$

where f is holomorphic and also $z'' = f(z')$ along V. The mapping (148) is always locally biholomorphic. To find the desired function f, we proceed as follows. The projection $\pi : \mathbb{C}^n \to \mathbb{C}^{q+k}$, when restricted to V, defines a CR diffeomorphism from V to $\pi(V)$. We denote the inverse CR mapping by g; its first $q + k$ components are just the first $q + k$ coordinates. We need to extend the last $n - (q + k)$ components to be holomorphic. To accomplish this, we first choose a real submanifold $S^{\#}$ of $\pi(V)$ such that $S^{\#}$ has no complex structure in its tangent spaces at all, and is of maximal dimension. Thus $S^{\#}$ has real dimension $q + k$. Its tangent spaces look like $(\mathbf{R} \times 0)^{q+k}$. The last components of g are now the restrictions of holomorphic functions f on the ambient space $\mathbb{C}^n$. These functions vanish on the manifold $S^{\#}$ by construction. It remains only to show that they vanish on $\pi(V)$. This is immediate from Lemma 5 below.

∎

LEMMA 5

Suppose that $S^{\#}$ is a totally real, real analytic submanifold of a CR manifold $\pi(V)$ of maximal dimension. Any CR function that vanishes on $S^{\#}$ also vanishes on $\pi(V)$.

PROOF   Consider a local basis of real vector fields for $\pi(V)$. These vector fields can be divided into three types: X_m, Y_m, D_j, where $1 \leq m \leq q$, $1 \leq j \leq k$, and where all the X_m, D_j are tangent to $S^{\#}$. Also the Y_m can be chosen so that $\{X_m - iY_m\}$ form a basis for $T^{1,0}(\pi(V))$. Then the CR equations are precisely the equations that

$$(X_m + iY_m)(f) = 0. \tag{149}$$

Since the $\{Y_m\}$ are linearly independent, they are dual to the differentials of functions y_j that can be taken as defining functions for $S^{\#}$. Now any real analytic function f vanishing there can be expanded into a series (without constant term) in the functions y_j:

$$f = \sum_{|J|=d}^{\infty} a_J(.)\, y^J. \tag{150}$$

In (150) we let d denote the order of vanishing of f in these variables, assuming that f does not vanish identically. (We ignore orders of vanishing in the other variables.) Apply any of the Cauchy–Riemann operators. From

$$(X_m + iY_m)(f) = 0 \tag{151}$$

and also $X_m(y) = 0$, we see that f vanishes to order both d and $d+1$ in the variables y_j. This is impossible unless d is infinite, in which case f vanishes identically. Note that this part of the argument generalizes one of the proofs that a holomorphic function in one variable that vanishes along the real axis must vanish identically. This finishes the proof of Lemma 4 and hence also that of Lemma 5. ∎

Observe that Lemma 4 says nothing for a hypersurface. For a hypersurface $q+k$ equals n. The lemma has content when $q+k$ is smaller than the dimension n. When $k = 0$, the result is also of some interest. The conclusion then is that the real $2q$-dimensional manifold can be fit into $\mathbb{C}^q$; hence it must be an open subset of $\mathbb{C}^q$. In particular, it is a complex manifold.

LEMMA 6
(The Frobenius theorem). *A subbundle of the tangent bundle of a real manifold is the tangent bundle of a submanifold if and only if it is integrable. (integrable means closed under the Lie bracket operation.)*

PROOF See for example [La]. ∎

We need one more lemma. Recall that we are dealing with germs.

LEMMA 7
Suppose that M is a real analytic pseudoconvex hypersurface and that $0 \in M \subset \mathbb{C}^N$. Suppose further that we have a CR submanifold $W \subset \mathbb{C}^{N-1} \times 0$ containing 0 such that the following hold:

1. *Let q denote the dimension of $T_p^{1,0}W$ for all points p near the origin. Assume that $q + k = N - 1$, and that*

$$T_0 W \simeq \mathbb{C}^q \times \mathbf{R}^k \times 0 \subset \mathbb{C}^q \times \mathbb{C}^{N-1-q} \times \mathbb{C}. \tag{152}$$

2. *Every (real) vector field on W is the real part of a (1,0) vector field on M that is annihilated by the Levi form.*

3. *There is an integer j so that the iterated brackets up to order j of vector fields in $T^{1,0}W$ and their conjugates span all of $\mathbb{C}TW$.*

Then $\mathbb{C}^{N-1} \times 0$ lies in M.

PROOF OF LEMMA 7   This result is Proposition 3 from [DF]. We sketch a similar proof. Assume that r is a defining function for M and that $r_0 : \mathbb{C}^{N-1} \to \mathbb{R}$ is defined by $r_0\left(z'\right) = r\left(z', 0\right)$. We wish to show that r_0 vanishes identically. The submanifold W is defined by the equations

$$z_N = 0$$

$$\rho_1 = \cdots = \rho_k = 0. \tag{153}$$

We may choose local holomorphic coordinates so that these defining equations have the form

$$\rho_j(z) = 2\,\mathrm{Re}\left(z_{q+j}\right) + \text{higher order terms}$$

$$j = 1, \ldots, k. \tag{154}$$

By the implicit function theorem we may also choose real local coordinates $s_1, \ldots, s_{2q+k}, \rho_1, \ldots, \rho_k$. We have a collection of real vector fields

$$\mathrm{Re}\left(L_j\right),\ \mathrm{Im}\left(L_j\right) \qquad j = 1, \ldots, q$$

$$D_m \qquad m = 1, \ldots, k \tag{155}$$

that annihilate the defining equations (154). We number these so that D_1 is an iterated commutator of the $\mathrm{Re}\left(L_j\right)$, $\mathrm{Im}\left(L_j\right)$; that D_2 is an iterated commutator of the $\mathrm{Re}\left(L_j\right)$, $\mathrm{Im}\left(L_j\right)$, D_1; and so forth.

The function r_0 vanishes to order at least two at every point of W, and we may suppose that the order of vanishing of this function is minimal at the origin. Using multi-index notation, write

$$\rho^J = \prod_{j=1}^{k} \rho_j^{J_j} \tag{156}$$

and expand r_0 in a Taylor series in these coordinates. Assuming that the Taylor expansion is not identically zero, we may write

$$r_0 = \sum_{|J|=m} A_J(s)\, \rho^J + O\left(|\rho|^{m+1}\right) \tag{157}$$

for m minimal. Computation of $\partial\bar{\partial} r_0$ shows that

$$\partial\bar{\partial} r_0 = \sum_{i,j=1}^{k} A_{Iji}(s)\, \rho^I \partial\rho_j \bar{\partial}\rho_i + O\left(|\rho|^{m-1}\right) \tag{158}$$

where the order of the multi-index I equals $m - 2$, and the functions A_{Iji} are constants times the original coefficients.

The lowest order term in (158) cannot possibly define a semidefinite form when m is odd, because it changes sign in every neighborhood of the origin. The form, however, must be semidefinite, by the pseudoconvexity of the hypersurface M. This proves that m cannot be odd.

It remains to show that the order of vanishing cannot be even. Again the idea is to contradict pseudoconvexity. For simplicity we give the details only when $k = 1$. The general case is similar, but the notation is more difficult. It appears in [DF]. See also Example 7, whose concreteness may clarify the ideas. We may suppose that the defining equation ρ_1 takes the form

$$\rho_1 = 2 \operatorname{Re}\left(z_{n-1}\right) + f\left(z', \overline{z'}, \operatorname{Im}\left(z_{n-1}\right)\right) \tag{159}$$

in appropriate coordinates. Here $z' = (z_1, \ldots, z_{n-2})$. Using the notation of (155), it follows from the third hypothesis that the vector field D_1 is obtained as an iterated commutator of the other vector fields. This condition amounts to saying that the function $f(z', \overline{z'}, 0)$ vanishes to finite order $d \geq 2$ at the origin. (Compare this with the results on finite type in $\mathbb{C}^2$ from Chapter 4.) After a linear coordinate change in the z' variables, we may suppose that $f(z', \overline{z'}, 0)$ includes the term $z_1^a \overline{z_1}^{d-a}$ in its Taylor expansion for some $a > 0$. Actually, by working at a generic point, we may in fact suppose that $d = 2$, and $D_1 = \operatorname{Im}\left[L, \overline{L}\right]$ for some L. This simplification is convenient but not necessary for the proof.

The idea again is to compute $\partial \overline{\partial} r_0$. Since we are assuming $k = 1$, we may divide by a unit so that the coefficient of ρ^m in (157) is unity. Computing $\partial \overline{\partial} r_0$ yields

$$
\begin{aligned}
\partial \overline{\partial}\left(\rho^m + A_{m+1}\rho^{m+1} + \cdots\right) &= \partial\left(m\rho^{m-1}\overline{\partial}\rho + (m+1)A_{m+1}\rho^m \overline{\partial}\rho + \cdots\right) \\
&= m(m-1)\rho^{m-2}\partial\rho\overline{\partial}\rho + m\rho^{m-1}\partial\overline{\partial}\rho \\
&\quad + m(m+1)A_{m+1}\rho^{m-1}\partial\rho\overline{\partial}\rho + \cdots. \tag{160}
\end{aligned}
$$

We must eliminate the lowest order term. To do so, we evaluate the last line in (160) on vectors L satisfying $\langle \partial\rho, L \rangle = 0$ along W. Then the terms of lowest order are of odd order $m - 1$ and equal

$$m\rho^{m-1}\left\langle \partial\overline{\partial}\rho, L \wedge \overline{L} \right\rangle. \tag{161}$$

Assuming that $\left\langle \partial\overline{\partial}\rho, L \wedge \overline{L} \right\rangle \neq 0$, (161) is again impossible because of the sign change. To be sure of this condition, we need only to choose L so that $\left\langle \overline{\partial}\rho, [L, \overline{L}] \right\rangle \neq 0$ everywhere along W. If this were to vanish everywhere, then the Levi form on W would vanish also. This would mean that $HW = T^{1,0}W \oplus T^{0,1}W$ would be integrable. This contradicts the existence of the vector field D_1. Thus $\langle \partial\bar{\partial}\rho, L \wedge \bar{L} \rangle \neq 0$ at points on W arbitrarily close to the origin. Since $m - 1$ is odd, we again contradict the pseudoconvexity of M.

Equivalently, we may restrict to an appropriate holomorphic curve z as in Chapters 3 and 4 and then compute the Laplacian

$$\frac{\partial^2 (z^* r_0)}{\partial t \partial \bar{t}} . \tag{162}$$

In case m is even, the condition $\langle \partial\rho, L \rangle = 0$ gets replaced by $[\partial (z^* \rho)]/\partial t = 0$.

The proof that the order cannot be even when k is larger than one is more difficult, but similar. One restricts to appropriate curves, and contradicts pseudoconvexity. This finishes the proof. ∎

We are now prepared to prove the important result of Diederich and Fornaess.

PROOF OF THEOREM 9   Let $M \subset \mathbb{C}^n$ be real analytic and pseudoconvex, and let $V \subset M$ be a real analytic subvariety of holomorphic dimension q. Let N denote the kernel of the Levi form. We start with an arbitrary point $p \in V$ and a neighborhood U_p of p in $\mathbb{C}^n$. We may pass to some regular point p_o in this neighborhood such that $\dim(T^{1,0}(V) \cap N)$ is constant (and equal to q) for all points near enough to p_o, and V is a manifold there. Let L_j, for $j = 1, \ldots, q$, denote linearly independent vector fields in $T^{1,0}(V) \cap N$. Consider the Lie algebra $A \subset TV$ generated by $\text{Re}(L_j)$, $\text{Im}(L_j)$ for $j = 1, \ldots, q$. After restricting to a perhaps smaller open set, where the rank is maximal, this is the bundle whose local sections are generated by all the iterated brackets of these real vector fields. By construction, it is integrable. By the Frobenius theorem we can therefore find a real analytic manifold V' for which $TV' = A$. This manifold still has holomorphic dimension at least q, because the L_j are tangent to V'. The manifold V' has the additional property that $TV' \subset hM$. The real bundle hM was defined in Section 3.1.2. Thus we have found a real analytic CR manifold of dimension $2q + k$. After making a change of coordinates as guaranteed by Lemma 4, we obtain a manifold that sits in $\mathbb{C}^{q+k} \times \{0\}$. If $q + k = n - 1$, then we have the hypotheses of Lemma 7 with $N = n$. More generally, we may have $q + k < n - 1$. By restricting all considerations to the subspace defined by setting $z^{q+k+1} = z^{q+k+2} = \cdots = z^{n-1} = 0$, we may replace n by $N = q + k + 1$. The restriction of the pseudoconvex hypersurface M to this subspace remains pseudoconvex. Thus we have reduced ourselves to the hypotheses of Lemma 7. It then follows from that lemma that the defining function r, restricted (locally) to $\mathbb{C}^{q+k} \times \{0\}$, vanishes identically along V'. Thus we have found the germ of a complex manifold of dimension at least q lying in the hypersurface M. ∎

We close this chapter with an example where $q = 1$ and $k = 2$.

Example 7

Put

$$\rho_1(z, \bar{z}) = 2\,\mathrm{Re}\,(z_1) + |z_2|^2$$
$$\rho_2(z, \bar{z}) = 2\,\mathrm{Re}\,(z_2) + |z_3|^2 \tag{163}$$

and define a real submanifold of $\mathbb{C}^4$ by

$$W = \{z : \rho_1(z, \bar{z}) = \rho_2(z, \bar{z}) = z_4 = 0\}. \tag{164}$$

Then W is a four-dimensional CR submanifold for which $q = 1$ and $k = 2$. Suppose that $W \subset M$, where $M \subset \mathbb{C}^4$ and L_1, L_2, L_3 form the usual local basis for $\mathbb{C}TM$. The vector field

$$L = \overline{z_2}\,\overline{z_3} L_1 - \overline{z_3} L_2 + L_3 \tag{165}$$

satisfies the following properties:

$$L \in T^{1,0}W$$
$$[L, \overline{L}] = B - \overline{B}$$
$$[[[L, \overline{L}], L], \overline{L}] = 2(L_1 - \overline{L_1}). \tag{166}$$

Here $B = (-\overline{z_2} + |z_3|^2)L_1 + L_2$. Note that neither it nor L_1 is tangent to W, although their imaginary parts are both tangent. Using the notation from Lemma 7 we have $q = 1$ and $k = 2$. According to that result, we can draw the following conclusion. If M is pseudoconvex and all of the local sections of TW are annihilated by the Levi form along W, then M contains $\mathbb{C}^3$. In this case it is easy to verify that a real analytic function vanishing to at least second order at the origin, and in the ideal generated by (ρ_1, ρ_2), cannot be plurisubharmonic unless it vanishes. To do so, restrict to a curve of the form $z(t) = (c_1 t^4 + c_2 t^5, c_3 t^2 + c_4 t^3, t)$, and compute the Laplacian (162). $\quad\square$

7

Analysis on Finite Type Domains

7.1 The Bergman projection

7.1.1 Definition of the Bergman projection and kernel function

Our concern remains the relationship between the geometry of the boundaries of domains and holomorphic mappings between them. One link between these ideas is the theory of the Bergman projection operator P. One passes from geometric information at the boundary to the $\bar{\partial}$-Neumann problem via the work of Kohn and Catlin. The theory of the $\bar{\partial}$-Neumann problem gives us information about the Bergman projection because of the formula

$$P = I - \bar{\partial}^* N \bar{\partial} \tag{1}$$

that relates P to the $\bar{\partial}$-Neumann operator on $(0,1)$ forms. The work of Bell shows how to prove results on boundary smoothness of biholomorphic and proper mappings when the projection operator satisfies a regularity property known as Condition R. Subelliptic estimates for the $\bar{\partial}$-Neumann problem guarantee that Condition R holds. We turn now to the theory of the Bergman projection.

Suppose that Ω is a bounded domain in $\mathbb{C}^n$. Let $L^2(\Omega)$ denote as usual the Hilbert space of square integrable (equivalence classes of) complex-valued functions on Ω. In this chapter we write $||f||_\Omega$ or $||f||_{L^2(\Omega)}$ for the usual norm when we need to specify the domain. We leave to the reader the verification that the square integrable holomorphic functions form a closed subspace of $L^2(\Omega)$. Let us denote this subspace by $H(\Omega)$.

DEFINITION 1 *The Bergman projection operator*

$$P : L_2(\Omega) \rightarrow H(\Omega) \tag{2}$$

is the orthogonal projection onto the closed subspace of square integrable holomorphic functions.

Until the work of Bell [Be2], most work in this area concentrated on the Bergman kernel function. The Bergman kernel function $K(z, w)$ is the integral kernel associated with the Bergman projection. Thus one has the formula

$$Pf(z) = \int_{\Omega} K(z, w) f(w) \, dV(w) \tag{3}$$

whenever f is square integrable. One of the main points of the work of Bell is a transformation law for the Bergman projections of two domains under a proper holomorphic mapping between them. In order to prove this formula we first need a lemma concerning extensions of square integrable holomorphic functions.

LEMMA 1
(L^2 removable singularities lemma). *Suppose that $\Omega \subset\subset \mathbb{C}^n$ and that $V \subset \Omega$ is a complex analytic subvariety of positive codimension. Then any square integrable holomorphic function on $\Omega - V$ extends to be holomorphic on all of Ω.*

PROOF We may first assume that the variety is of codimension one, since the larger the codimension the smaller the variety. As it is enough to give a local extension of the function across the variety, we may also assume without loss of generality that we are in the following situation. The domain Ω is the unit polydisc about the origin, and the variety is defined by the vanishing of the coordinate function z_n. Write $z = (z_1, \ldots, z_{n-1}, z_n) = (w, t)$ where $z_n = t$, and denote the polydisc by $A \times B$. We may expand any holomorphic function f in $\{(w, t) : |w| < 1, 0 < |t| < 1\} = A \times (B - 0)$ into a Laurent series in t with holomorphic coefficients in w. We write this as

$$f(w, t) = \sum_{-\infty}^{\infty} f_m(w) t^m. \tag{4}$$

Using the orthogonality of the powers of t, we see easily that

$$\|f\|_{A \times B}^2 = \sum_{-\infty}^{\infty} \|f_m\|_A^2 \, \|t^m\|_{B-0}^2. \tag{5}$$

Since the sum in (5) is convergent, the coefficients must vanish when the index of summation is negative. Thus the Laurent series defines in fact a holomorphic function in the polydisc. This gives the required extension. ∎

We can now state and prove the transformation formula for the Bergman projection.

PROPOSITION 1

(Transformation law). *Suppose that* $f : \Omega_1 \rightarrow \Omega_2$ *is a proper holomorphic mapping between bounded domains in* $\mathbb{C}^n$. *Let* P_1, P_2 *denote the corresponding Bergman projection operators. Then the following transformation law holds:*

$$P_1 \left(\det(df) \left(\phi(f) \right) \right) = \det(df) \left((P_2 \phi)(f) \right) \tag{6}$$

for every ϕ *in* $L^2(\Omega_2)$.

PROOF First, the proper map $f : \Omega_1 \rightarrow \Omega_2$ has a global multiplicity m. This means that it is an m-sheeted covering map $f : \Omega_1 - V_1 \rightarrow \Omega_2 - V_2$ between the complements of the branching loci. The branch locus in Ω_1 is the variety defined by the vanishing of $\det(df)$, and the proper image of a complex analytic variety is itself a complex variety. These loci are complex analytic subvarieties of codimension one, and hence have measure zero. According to the change of variables formula, and using Lemma 1.2 yet again, we have

$$\int_{\Omega_1 - V_1} |\det(df)|^2 |\phi(f)|^2 \, dV = m \int_{\Omega_2 - V_2} |\phi|^2 \, dV \tag{7}$$

Since the varieties V_i have measure zero, we obtain that

$$\|\det(df)(\phi(f))\|_{L^2(\Omega_1)} = \sqrt{m} \, \|\phi\|_{L^2(\Omega_2)} . \tag{8}$$

From (8) it follows that $g = \det(df)(\phi(f)) \in L^2(\Omega_1)$ in case $\phi \in L^2(\Omega_2)$. Next we wish to write $\langle g, \psi \rangle_{\Omega_2} = \langle \phi, h \rangle_{\Omega_1}$ for an appropriate $h \in L^2(\Omega_2)$. The right choice is

$$h = \sum_{k=1}^{m} \det(dF_k)(\psi(F_k)) \tag{9}$$

where the functions F_k are the local inverses of f. Note that h is just (m-times) the symmetrized version of $\det\left(df^{-1}\right)\left(\psi\left(f^{-1}\right)\right)$ and is thus well defined. A simple estimate shows that h lies in $L^2(\Omega_2 - V_2)$, so the inner products make sense. Let us apply this formula when the function ψ is also holomorphic. Then the function h is holomorphic on $\Omega_2 - V_2$ and is in $L^2(\Omega_2 - V_2)$. According to the L^2 removable singularities lemma, it extends to be in $H(\Omega_2)$. This yields

$$\langle \det(df)(\phi(f)), \psi \rangle_{\Omega_1} = \langle g, \psi \rangle_{\Omega_1} = \langle \phi, h \rangle_{\Omega_2}$$
$$= \langle \phi, P_2 h \rangle_{\Omega_2} = \langle P_2 \phi, h \rangle_{\Omega_2} . \tag{10}$$

Now apply the change of variables formula to the last inner product, obtaining

$$
\begin{aligned}
\langle P_2\phi, h\rangle_{\Omega_2} &= \int_{\Omega_2} P_2\phi\,\overline{h}\,dV \\
&= \int_{\Omega_1} (P_2\phi)(f)\overline{(h(f))}\,\det(df)\,dV \\
&= \langle \det(df)(P_2\phi)(f), \psi\rangle_{\Omega_1}.
\end{aligned}
\tag{11}
$$

This shows that $\det(df)(P_2\phi)(f)$ and $\det(df)(\phi(f))$ have the same inner product with any holomorphic ψ. Therefore they have the same Bergman projections, namely,

$$
\begin{aligned}
P_1(\det(df)(\phi(f))) &= P_1(\det(df)(P_2\phi)(f)) \\
&= (\det(df)(P_2\phi)(f)).
\end{aligned}
\tag{12}
$$

This is the formula we are trying to prove. Note that we have used in the last step the fact that $\det(df)(P_2\phi)(f)$ is holomorphic. ∎

7.1.2 Condition R

In this section we discuss the regularity property Condition R of the Bergman projection operator and its connection to the Neumann operator. Let Ω be a smoothly bounded pseudoconvex domain. Suppose additionally that each boundary point of Ω is a point of finite one-type. Then, by the result of Catlin, there is a subelliptic estimate on $(0,1)$ forms. The link between the Neumann operator and the Bergman projection follows from the formal properties of the $\overline{\partial}$-Neumann problem; we now prove the formula of Kohn that relates them:

LEMMA 2
(Kohn). *The Bergman projection P can be expressed in terms of the $\overline{\partial}$–Neumann operator as follows:*

$$
P = I - \overline{\partial}^* N \overline{\partial}
\tag{13}
$$

PROOF If $f \in L^2(\Omega) \cap O(\Omega) = H(\Omega)$, then $\overline{\partial}f = 0$ and $Pf = f$, so the formula is correct. If, on the other hand, $f \in H(\Omega)^\perp$, then we solve the Cauchy–Riemann equation

$$
\overline{\partial}u = \overline{\partial}f
\tag{14}
$$

for the unique solution, namely f, orthogonal to the holomorphic functions.

According to the formalism of the $\overline{\partial}$-Neumann problem, we must have

$$f = \overline{\partial}^* N \left(\overline{\partial} f \right) \qquad (15)$$

Again the formula holds. Thus its truth is established on complementary subspaces, and hence on all of $L^2(\Omega)$. ∎

A subelliptic estimate implies that N is "globally regular." From Theorem 6.4 and Lemma 2 therefore, we obtain Condition R whenever subelliptic estimates hold at every boundary point. Because of its importance we redundantly state this particular conclusion as a theorem.

THEOREM 1
A smoothly bounded pseudoconvex domain for which each boundary point is a point of finite type satisfies Condition R.

The precise definition of global regularity is as follows. The $\overline{\partial}$-Neumann problem on (0,q) forms is globally regular if N maps the space of forms of type (0,q) whose coefficients are in $C^\infty(\overline{\Omega})$ to itself. There are general situations when N satisfies this property. Boas [Bo] proved this on smoothly bounded pseudoconvex domains for which the set of points of infinite type is small, for example of Hausdorff two-dimensional measure zero. Boas and Straube [BS2] proved this on any smoothly bounded domain admitting a defining function that is plurisubharmonic on the boundary. (This holds for example on convex domains.) The same authors [BS1] determine the precise relationship between the regularity properties of the Neumann operators and Bergman projection operators. There it is important to consider these operators on forms of different degrees. In particular, N is globally regular on (0,1) forms if and only if the usual Bergman projection and the corresponding projections onto $\overline{\partial}$-closed (0,1) and (0,2) forms are all globally regular. For our purposes we need the regularity only for the usual Bergman projection. Bell had given earlier the name "Condition R" to this regularity property.

DEFINITION 2
Suppose that Ω is a smoothly bounded domain in C^n. Then Condition R holds on Ω if the Bergman projection P maps $C^\infty(\overline{\Omega})$ to itself.

It seems to be very difficult to determine precisely what geometric conditions on a domain are necessary and sufficient for Condition R to hold. Barrett [Ba3] has constructed domains in $\mathbb{C}^2$ that are not pseudoconvex and for which Condition R fails. Kiselman [Ki] has shown that there are bounded pseudoconvex domains with nonsmooth boundaries in $\mathbb{C}^2$ for which Condition R fails. He constructs additional examples, including a smoothly bounded pseudoconvex domain in a two-dimensional complex manifold, for which Condition R fails.

By the Boas–Straube results cited above, we know that Condition R holds on some classes of domains more general than domains of finite type. For the class of domains of finite type, the proof that Condition R holds requires at present the machinery of the $\bar{\partial}$-Neumann problem and Catlin's theorem on subelliptic estimates. Thus establishing this condition seems to be very deep. Its usefulness lies in proving boundary regularity theorems for biholomorphic and proper mappings. In the next section we indicate how to prove these extension results given that Condition R holds.

7.1.3 Smooth extension to the boundary for proper mappings

There are several methods for obtaining holomorphic extensions of holomorphic mappings. Any of these methods can be used to prove results on boundary smoothness. In particular, if Ω is a domain with smooth boundary, and

$$f \in O(\Omega) \tag{16}$$

has a holomorphic extension to a larger domain, then we automatically obtain a smooth extension to (a subset of) $b\Omega$. On the other hand, it is not generally true that a proper holomorphic mapping between given smoothly bounded domains has a holomorphic extension to a larger domain. Thus many authors have developed techniques for proving smooth extension results (see [Fo3]). The most successful methods to date have been the use of the theory of the Bergman kernel function by Fefferman [Fe] and the use of the Bergman projection by Bell. These approaches require the boundedness of the domains. Before turning to the results themselves, let us give a simple example that illustrates why restricting to bounded domains is reasonable.

Example 1
Put

$$\Omega_1 = \left\{ (z_1, z_2) : |z_1|^2 + |z_2|^2 < 1 \right\}$$

$$\Omega_2 = \left\{ (w_1, w_2) : |w_1 (1 - w_2)|^2 + |w_2|^2 < 1 \right\}$$

$$f(z_1, z_2) = \left(\frac{z_1}{1 - z_2}, z_2 \right) = (w_1, w_2). \tag{17}$$

Then $f : \Omega_1 \to \Omega_2$ is a biholomorphic mapping from the ball to an unbounded domain. It is obviously not smooth at the boundary point $(0, 1)$. Note that $b\Omega_2$ contains the one-dimensional complex line given by $w_2 = 1$. If we consider f^{-1}, then we have a biholomorphic mapping from a domain whose boundary contains a complex analytic variety to the ball. This shows that the restriction to bounded domains in Theorem 2 is necessary. This example reveals also that one must assume boundedness to have a reasonable chance for smooth extension to the boundary.   []

THEOREM 2

(Bell). *Suppose that Ω is a smoothly bounded pseudoconvex domain for which Condition R holds. Then any proper holomorphic mapping from Ω to another smoothly bounded pseudoconvex domain (in the same dimension) extends to be a smooth mapping of the boundaries.*

We omit the proof and refer the reader to [Be2,Be3]. There are many consequences of Theorem 2. Note in particular that the conclusion follows also for biholomorphic mappings. By applying the theorem also for the inverse mapping, we have the following famous corollary. The first proof, due to Fefferman, applied only to strongly pseudoconvex domains, and required a deep analysis of the Bergman kernel function on a strongly pseudoconvex domain. Bell took the significant step of considering the Bergman projection operator rather than the Bergman kernel function as the basic object. Other authors such as Webster and Ligocka also made important simplifications. Today there are simpler proofs in the strongly pseudoconvex case, but these do not seem to apply more generally (see [Fo3]).

COROLLARY 1

Suppose that $f : \Omega_1 \to \Omega_2$ is a biholomorphic mapping of smoothly bounded domains in the same dimension, and each domain satisfies Condition R. Then f extends to be a smooth diffeomorphism of the boundaries.

COROLLARY 2

Suppose that Ω_1, Ω_2 are smoothly bounded domains in $\mathbb{C}^n$ that each satisfy Condition R. Suppose that $p \in b\Omega_1$, and $G(p)$ is any differential geometric invariant of the boundary. If $f : \Omega_1 \to \Omega_2$ is biholomorphic, then $G(p) = G(f(p))$. This applies in particular in case $G(p)$ is any of the numbers described in Theorem 4.11.

Example 1 shows that the conclusion fails in general for unbounded domains in $\mathbb{C}^n$. It is surprising that the conclusion fails for bounded domains in complex manifolds. Barrett [Ba2] has given examples of smoothly bounded biholomorphically equivalent domains in complex manifolds for which the boundaries are not CR equivalent. In fact, his examples show that the type of a boundary point need not be preserved. These examples show that without some restrictions on the underlying complex manifold, it is unreasonable to expect that boundary invariants of hypersurfaces are biholomorphic invariants of the domains they bound.

7.2 Boundary invariants and CR mappings

7.2.1 An application

Let us return to $\mathbb{C}^n$. To illustrate the power of the result on smooth extendability of proper mappings, we show that there are no proper mappings between certain domains. These examples are natural generalizations of the approach taken in Chapter 1. There we showed that there was no proper mapping from a polydisk to a ball because the sphere did not contain any complex analytic varieties. Now we can show that there are no proper mappings between smoothly bounded pseudoconvex domains unless the orders of contact of complex analytic varieties with the boundaries match up correctly. More generally other geometric invariants must also correspond. For example, to each point in the boundary we have assigned ideals of holomorphic functions. These must also transform correctly in order that there be a proper mapping between the domains. We indicate this with some examples and statements of theorems. To simplify the exposition, we assume that the boundary is real analytic.

We state an important result in the real analytic case due to Baouendi and Rothschild [BR1] in Theorem 3 below. The proof of this result relies strongly on the smooth extendability to the boundary of the proper mapping. The hypothesis that the (compact) boundaries are real analytic guarantees by the Diederich–Fornaess theorem (Theorem 4.5) that the domains contain no complex analytic varieties and hence by Theorem 4.4 that each boundary point is a point of finite type. Thus, either by the general result of Catlin or the result of Kohn in the real analytic case, there is a subelliptic estimate at each boundary point. By the results of Bell from this chapter, Condition R holds on each domain, and hence proper holomorphic mappings extend smoothly to the boundary. Once we have the smooth extension, the methods of Baouendi and Rothschild show how to get a holomorphic extension past the boundary. This yields the following result.

THEOREM 3
Suppose that $f : \Omega_1 \to \Omega_2$ is a proper holomorphic mapping between bounded pseudoconvex domains with real analytic boundary. Then f extends holomorphically past $b\Omega_1$.

The extension will be a finite holomorphic mapping in some neighborhood of any boundary point. Recall from the work in Chapter 2 that multiplicities multiply under finite holomorphic mappings. There we verified that if $f : (\mathbb{C}^n, 0) \to (\mathbb{C}^n, 0)$ is a finite holomorphic map germ, and if $I \subset {}_n O$ denotes any ideal in the range space, then

$$\mathbf{D}\left(f^* I\right) = \mathbf{D}(f)\,\mathbf{D}(I). \tag{18}$$

By combining these results, we obtain necessary conditions that there be a proper holomorphic mapping between domains.

Consider the following simple example.

Example 2
Put

$$\Omega_p = \left\{ z : \sum_{j=1}^{n} |z_j|^{2p_j} < 1 \right\} \tag{19}$$

for an n-tuple of positive integers $p = (p_1, \ldots, p_n)$. We may suppose that $p_1 \geq p_2 \geq \cdots \geq p_n$. Then there can be no proper mapping from $f : \Omega_p \to \Omega_s$ unless $p_j \geq s_j$ for each j. In fact s_j must divide p_j. There are many ways to prove this, and in fact in this simple example, one doesn't need to know that the mapping extends smoothly to the boundary to obtain the result. More generally we could consider domains defined by

$$r(z, \bar{z}) = 2 \operatorname{Re}(z^n) + |z^n|^2 + \sum_{j=1}^{n-1} |h_j(z)|^2$$

$$s(z, \bar{z}) = 2 \operatorname{Re}(z^n) + |z^n|^2 + \sum_{j=1}^{n-1} |g_j(z)|^2 \tag{20}$$

where the functions are polynomials. We wish to decide whether there is a proper mapping f from the domain $\{r < 0\}$ to the domain $\{s < 0\}$; we make the assumptions that

$$\mathbf{D}\left(h_1, h_2, \ldots, h_{n-1}, z^n\right) < \infty$$

$$\mathbf{D}\left(g_1, g_2, \ldots, g_{n-1}, z^n\right) < \infty. \tag{21}$$

Then each domain is bounded, pseudoconvex, and of finite type. By the Baouendi-Rothschild result a proper mapping f between them extends holomorphically past the origin, and is a finite mapping there. It then follows easily that $f_n(z) = z_n$. We then need to check only whether $g = fh$. One can often see that this fails by using some of the algebraic methods of Chapter 2; in particular we must have $\mathbf{D}(g) \geq \mathbf{D}(h)$. ☐

Suppose that $\Omega \subset\subset \mathbb{C}^n$ is a pseudoconvex domain with smooth boundary M. Let p be a boundary point, and recall that $I(U, p, k)$ is the holomorphic family of ideals associated with the kth order osculation of the boundary by an algebraic hypersurface. Choose k much larger than the type at p. We can in principle decide whether there is a proper mapping between Ω and another domain by seeing how the associated families of holomorphic ideals transform under proper mappings.

7.2.2 CR mappings

Suppose that Ω is a bounded domain in $\mathbb{C}^n$ with smooth boundary. In this book we have associated ideals of holomorphic functions with each boundary point. Under favorable circumstances, these ideals are biholomorphic invariants of the domain itself. In order to prove such a statement, one needs to know that a biholomorphic mapping of Ω extends to be a diffeomorphism of its boundary. As we have seen, much research in several complex variables since 1974 has been devoted to this question and its generalizations. One can turn the question around. Suppose that $f : M \rightarrow M'$ is a (smooth, or perhaps just continuous) CR mapping of smooth real hypersurfaces. Can one extend f (locally) to be a holomorphic mapping of one or perhaps both sides of M? This question has also motivated much closely related research. We refer to [Bg] for the latest results. We do state the following necessary and sufficient condition for extendability of a CR mapping to at least one side of a hypersurface.

THEOREM 4

(Trepreau). *Suppose that M is a smooth real hypersurface in $\mathbb{C}^n$ and that $p \in M$ is an arbitrary point. Then every germ at p of a CR function on M extends (locally) to be holomorphic on one side of M if and only if there is no germ of a complex analytic hypersurface passing through p and lying in M.*

Note the difference in the condition of Theorem 4 from the condition of finite $(n - 1)$-type. These conditions are equivalent in the real analytic case; the extension result in that case was proved earlier by Baouendi and Treves [BT]. The extension to the pseudoconvex side of a CR function on a strongly pseudoconvex hypersurface is classical.

7.2.3 Regularity of CR mappings

In the final section we discuss a result of Bell and Catlin on the regularity of CR mappings. The idea is that a continuous CR mapping $f : M_1 \rightarrow M_2$ between pseudoconvex hypersurfaces of finite type that is finite-to-one must be smooth. If the mapping is also a local homeomorphism, then it must be in fact a local diffeomorphism. Thus the type of a point and, more generally, the constructions of Chapter 4 are invariants of the CR structure. This sort of result is proved by showing that the mapping has a holomorphic extension to the pseudoconvex side in such a way that the resulting mapping is smooth. In case the hypersurfaces are pseudoconvex and real analytic, the mapping extends to be holomorphic in an open set intersecting both sides of the hypersurface.

In order to prove these results one applies nearly all the ideas of this book; we omit the details. One requires the following local extendability theorem for proper holomorphic mappings.

THEOREM 5
Suppose that

$$\Omega_1, \Omega_2 \subset\subset \mathbb{C}^n \qquad (22)$$

are bounded pseudoconvex domains and $f : \Omega_1 \to \Omega_2$ *is a proper holomorphic mapping between them. Let* $z_0 \in b\Omega_1, w_0 = f(z_0) \in b\Omega_2$ *be points and suppose that there are neighborhoods* U_1, U_2 *of these points such that*

$$U_1 \cap \Omega_1 = M_1$$

$$U_2 \cap \Omega_2 = M_2 \qquad (23)$$

define smooth pseudoconvex (connected) hypersurfaces. Suppose that f *extends to be continuous up to the boundary and* $f(M_1) \subset M_2$. *Finally we suppose that* z_0 *is a point of finite type, and* $f^{-1}(w_0)$ *is a compact subset of* M_1. *Then* f *extends to be of class* C^∞ *near* $f^{-1}(w_0)$. *It is also necessarily finite-to-one near this set.*

PROOF The first step is to show that there is an estimate relating the distance between the image $f(z)$ and M_2 to the distance between z and M_1. This is the statement that there are positive constants c, t such that

$$\operatorname{dist}(f(z), M_2)^t \le c \operatorname{dist}(z, M_1) \qquad (24)$$

for z near z_0. This is a consequence of the compactness of $f^{-1}(w_0)$ and the fact that we can find a local smooth defining function ρ such that $-(-\rho)^t$ is plurisubharmonic on the pseudoconvex side of M_1. Once we know that such a function exists, we may form the continuous plurisubharmonic function $\rho^\#(w) = \sup\{-(-\rho(w))^t : w = f(z)\}$ and apply the Hopf lemma to it. Recall that the Hopf lemma enables us to estimate that $\rho^\#(w) \le -c \operatorname{dist}(w, M_2)$ for w close enough to w_0 and still in the interior. The estimate (24) now follows by comparing (minus) the defining function with the distance to the boundary.

Next one applies the techniques of Bell on the Bergman projection. One of the key lemmas is the following. Given any monomial z^a, we can find a function ϕ_a in $C^\infty(\Omega_2)$ that vanishes to infinite order along M_2 yet satisfies the equation

$$P_2(\phi_a) = z^a. \qquad (25)$$

According to the transformation law for the Bergman projection under proper holomorphic mappings, we have

$$\det(df)(f)^a = P_1\{\det(df)(\phi(f))\}. \qquad (26)$$

Recall that z_0 is a point of finite type, so there is a subelliptic estimate for the $\bar{\partial}$-Neumann problem. From this we see that Condition R holds. By applying the Cauchy estimates we obtain that the right-hand side of (26) is smooth up to the boundary. We must then "divide out" the determinant to conclude that the components of the mapping are themselves smooth up to the boundary. This point is elaborated in [BC1]; the point is a division theorem that applies whenever it is known that the determinant $\det(df)$ vanishes to finite order. ∎

In conclusion we observe that the behavior of analytic objects on domains in complex space is intimately connected to the geometry of the boundaries, and that one way to describe the geometry of the boundary is the approach outlined in this book. Questions about holomorphic mappings between bounded domains have led us to investigate how closely holomorphic data can touch the boundaries. In this book we have seen how to answer this using some basic algebraic geometry. The author hopes that the point of view herein will be useful in future considerations in complex analysis. In particular it seems possible that general results concerning the shape of approach regions through which holomorphic functions have boundary limits can be attacked by use of these ideas.

Problems

This collection of problems is divided into two parts. Many of the exercises in the first part do little more than test whether the reader understands the notation. Some of these problems come from the text. There are two problems concerning explicit computation of the Bergman kernel function. These are long exercises in advanced calculus. The problems in the second part have a different flavor. A star ($\star$) denotes that the problem is, so far as the author is aware, open. Even if no star appears, a problem in Part II will be difficult.

I. Routine problems

1. Suppose that p, q are polynomials in the two variables $t, \bar{t}$ but their product is a function of t alone. Then, if both are not identically zero, prove that each is just a function of t alone. Show that the conclusion fails for real analytic functions. Explain.

2. Show that a homogeneous polynomial of degree m is a linear combination of mth powers of linear functions.

3. Suppose that f is holomorphic on the unit ball in $\mathbb{C}^2$ and that $|f(z)| < 1/1 - ||z||$. What is the best estimate you can give for any kth derivative $D^k f(0)$? (Hint: Use calculus on the Cauchy estimates.)

4. Prove the uniqueness statements in the Weierstrass preparation and division theorems.

5. Give a formal proof of the Weierstrass preparation theorem that applies in the formal power series category.

6. Prove that the decomposition of a germ of a complex analytic subvariety into irreducible branches is unique up to order.

7. Prove the formal properties about ideals and germs of varieties:

$$\mathrm{rad}\,(I \cap J) = \mathrm{rad}\,(I) \cap \mathrm{rad}\,(J)$$

$$\mathbf{V}\,(I \cap J) = \mathbf{V}\,(I) \cup \mathbf{V}\,(J)$$

$$\mathbf{I}\,(V_1 \cup V_2) = \mathbf{I}\,(V_1) \cap \mathbf{I}\,(V_2)$$

$$\mathbf{V}\,(I) = \mathbf{V}\,(\mathrm{rad}\,(I)) .$$

8. Show that a prime ideal is a radical ideal, and give an example to show that the converse is false.

9. Determine the condition on the integers p, q so that the following ideal in O is prime:

$$\left(z_3^2 - z_1^p, z_2^3 - z_1^q \right) .$$

What is the maximum possible number of branches through the origin?

10. Compute the numbers $\mathbf{D}\,(I), \mathbf{T}\,(I), \mathbf{K}\,(I)$ if I is the ideal generated by each of the following:

$$\left(z_1^{p_1}, z_2^{p_2}, \ldots, z_n^{p_n} \right)$$

$$\left(z_1^3 - z_1 z_2 z_3, z_2^3 - z_1^3, z_3^3 \right)$$

$$\left(z_1^3 - z_2 z_3, z_2^4 - z_3^5, z_1^7 \right) .$$

11. Compute the intersection number $\mathbf{D}\,(f) = \dim_{\mathbf{C}} \left(O / (f) \right)$ of each of the ideals defined below by using any valid method:

$$\mathbf{D}\left(ab - c^3, ac - b^2, a^4 \right)$$

$$\mathbf{D}\left(a^3 - bcd, b^3 - acd, c^3 - abd, d^3 \right)$$

$$\mathbf{D}\left(ab - c^5, a^3 - bc^3, b^3 - ac^3 \right) .$$

12. Put

$$f\,(z, w) = \left(z^i - w^j, z^k w^l \right) .$$

Write

$$\begin{pmatrix} f_1 \\ f_2 \end{pmatrix} = \begin{pmatrix} a_{11} & a_{12} \\ a_{21} & a_{22} \end{pmatrix} \begin{pmatrix} z \\ w \end{pmatrix}$$

for appropriate functions inside the matrix. Verify that

$$\det\,(df) \equiv m \,\,(a_{11}a_{22} - a_{12}a_{21}) \bmod (f)$$

where

$$m = \mathbf{D}\,(f) .$$

Try to prove the appropriate generalization of this, that is,

$$\det(df) \equiv \mathbf{D}(f)\det(A) \bmod (f)$$

when $f(z) = A(z)z$ and f is a finite analytic map. The only way the author knows how to do this uses integrals and the statement that

$$h \equiv 0 \bmod (f) \Leftrightarrow \int \frac{hg\det(df)}{f}\, dz = 0 \quad \forall g.$$

13. Explicitly list a basis for $O/(f)$ in case

$$(f) = \left(z^3 - w^4, z^5 - w^5\right).$$

Verify in this example that the formulas (from Theorem 2.3)

$$\sum m_j(f_1)\, v\left(z_j^* f_2\right) = 15$$

$$\sum m_j(f_2)\, v\left(z_j^* f_1\right) = 15$$

hold where the lengths m_j are equal to unity and the sum represents the decomposition into irreducible branches. Do the same problem for the ideal

$$(f) = \left(z^3 - w^4, (z - w)^5\right)$$

and explain the differences.

14. Suppose that $M \subset \mathbb{C}^n$ is a real analytic hypersurface, defined by $r(z, \bar{z}) = 0$. Suppose there is an analytic identity of the form $H(z, \bar{z}) = 0$ for all $z \in M$. Prove that

$$H(z, w) = 0 \quad \text{whenever} \quad r(z, w) = 0.$$

15. Perform the computations needed to prove Proposition 3.1.

16. Compute the Levi form for hypersurfaces in $\mathbb{C}^n$ defined by $\sum_{j=1}^{N}|f_j(z)|^2 + 2\,\mathrm{Re}(z_n) = 0$ and find its determinant. Under what conditions is the origin a point of finite type? Assume $f_j(0) = 0$. More generally, find the determinant of the Levi form on a real analytic hypersurface defined by

$$2\,\mathrm{Re}(z_n) + \sum_{j=1}^{N}|f_j(z')|^2 - \sum_{k=1}^{N'}|g_j(z')|^2.$$

Discuss also the case where the sums are infinite.

17. Find an orthonormal basis $\{\phi_m\}$ for $L_2(\Omega) \cap O(\Omega)$ for domains of the form

$$\sum_{j=1}^{k}|z_j|^{2p_j} < 1.$$

Try to compute the Bergman kernel function $\sum_{m=1}^{\infty} |\phi_m(z)|^2 = K(z,z)$ for this domain. Under what circumstances can this function be represented as an elementary function?

18. Same problem for the domain

$$\left(\sum_{j=1}^{m} |z_j|^2\right)^P + \left(\sum_{j=m+1}^{n} |z_j|^2\right)^Q < 1.$$

Find a closed form expression for the kernel function when $Q = 1$.

19. Use the methods of Chapter 2 to generalize Rouche's theorem to (vector-valued) mappings in several variables.

20. (Reznick). Use the method of generating functions to show that the numbers

$$c_{rs} = \left(\frac{1}{4}\right)^{r-s} \sum_{k=s}^{r} \binom{2r+1}{2k} \binom{k}{s}$$

that arise in Chapter 5 are also equal to the simpler expression

$$c_{rs} = \binom{2r-s}{s-1} + \binom{2r-s+1}{s}.$$

21. Verify that a Blaschke product with exactly two factors is spherically equivalent to the map $z \to z^2$. What is the correct generalization to several variables?

22. Find, for each m, a proper holomorphic map from $f : B_2 \to B_{m+1}$ that is homogeneous of degree m. Think of this map as a map to some variety in $\mathbb{C}^{m+1}$. Find the equations for this variety when $m = 1, 2, 3, 4$. Prove that the origin is an isolated singular point. Show that, unless $m = 1$, the topology near 0 is not locally Euclidean.

23. Suppose that λ is a semidefinite form whose matrix entries are smooth functions. Show that λ can be smoothly diagonalized in a neighborhood of any point where there is at most one vanishing eigenvalue. Give an example of a λ that cannot be smoothly diagonalized.

24. Give an example of an ideal $I \subset O$ so that I is not equal to the ideal generated by the Weierstrass polynomials defined in (126) of Chapter 1. Show that, if the discriminant is a constant, then these ideals are the same.

25. Show that the ideal $(z_1^8 - z_3^5, z_1^7 - z_2^5)$ in O is not prime. Find all the branches of the variety through the origin. Give an example of a function that vanishes on a branch of the variety that is not in the ideal. Find a prime ideal defining the curve (t^5, t^7, t^8).

26. Verify that the automorphism group of the ball has the description given in Section 5.1.3. First verify that the maps ξ_a are in fact automorphisms. Use them to reduce to the case where the automorphism preserves the origin.

Give an analogue of the Schwarz lemma to prove that automorphisms of balls preserving the origin must be unitary.

27. For the finite unitary groups $\Gamma(p, q)$ defined in Section 5.2.2, exhibit a basis for the algebra of holomorphic functions invariant under the group.

28. Verify the assertion in Example 3.2 that the ideals there never define a positive-dimensional variety.

29. Put

$$r(z, \bar{z}) = 2 \, \text{Re} \, (z_3) + |z_1|^2 + |z_2 z_1|^2 + |z_2|^2 \, 2 \, \text{Re} \, (z_1^m) \, .$$

Compute the associated ideals. Find a complex analytic variety through the origin and lying in the hypersurface defined by r. Show that the hypersurface is essentially finite at the origin. Show that the hypersurface is pseudoconvex near the origin for appropriate m.

II. Harder and open problems

A star ($\star$) denotes an open problem.

$\star 1$. Let I_k denote the Kohn ideals of subelliptic multipliers. Prove that $1 \in I_k$ for some k holds at a point of finite type on a smooth pseudoconvex boundary.

$\star 2$. Give a formal power series proof of Catlin's theorem on subellipticity using Kohn's method of subelliptic multipliers.

$\star 3$. Prove using commutative algebra the fact that the Kohn ideals (in the holomorphic category) arising from the ideal $J_0 = (f_1, \ldots, f_N)$ eventually define the whole ring if and only if the variety of the ideal is trivial. Extend the proof to formal power series.

$\star 4$. Suppose $f : B_n \to B_N$ is a rational proper map. What is the maximum degree possible as a function of n, N.

$\star 5$. Give a different proof of the result in Chapter 5 that determines for which finite unitary groups of the form

$$\begin{pmatrix} \epsilon & 0 \\ 0 & \epsilon^q \end{pmatrix} \quad \epsilon^p = 1$$

there is a proper rational map from $B_2 \to B_N$ invariant under this group. Is there a differential-topological explanation?

⋆**6.** How smooth must a proper map $f : B_n \to B_N$ be to ensure that it is rational? Assume $n > 1$. In particular, must a C^1 proper map $f : B_n \to B_N$ be rational if $n > 1$?

7. Write a computer program to find all monomial proper maps $f : B_n \to B_N$ if $n > 1$ and $N > 2n - 2$ are fixed.

⋆**8.** Give a clean factorization of a rational proper map of the form

$$\frac{p_d + p_{d+1} + \cdots + p_{d+m}}{1 + q_1 + \cdots + q_m}.$$

⋆**9.** Characterize those domains that can be globally defined by $\sum_{j=1}^{\infty} |f_j|^2 < 1$ for holomorphic functions in the convergent sum.

⋆**10.** Describe precisely the boundary behavior of the Bergman kernel on a domain defined by $\sum_{j=1}^{N} |f_j|^2 < 1$ for holomorphic functions defining a trivial variety.

⋆**11.** Determine whether a point of finite type is a peak point for the algebra of holomorphic functions on a weakly pseudoconvex domain.

⋆**12.** Figure out a generalization of the Diederich–Fornaess theorem on varieties of positive holomorphic dimension that makes sense on a pseudoconvex CR manifold and prove it.

⋆**13.** Prove or disprove the statement that, on a pseudoconvex CR manifold, the statement that $c_p(L) = \text{type}(L, p)$ holds for all vector fields L of type $(1,0)$. Relate this to orders of contact to complex manifolds in case the CR manifold lives in some $\mathbb{C}^n$.

14. (McNeal). Suppose that $\Omega \subset \mathbb{C}^n$ is convex and has smooth boundary. Prove that $\Delta^1(b\Omega, p) = \Delta^1_{\text{reg}}(b\Omega, p)$. Give a generalization for $\Delta^q(b\Omega, p)$.

15. More generally, suppose that

$$r(z) = 2 \operatorname{Re}(z_n) + \|f\|^2 - \|g\|^2$$

is a real analytic defining function. Give a condition on f, g that is necessary and sufficient for $\Delta^1(b\Omega, p) = \Delta^1_{\text{reg}}(b\Omega, p)$ to hold?

⋆**16.** Determine necessary and sufficient conditions for Condition R.

⋆**17.** Give a reasonable criterion on a proper rational map between balls that is necessary and sufficient for the X-variety to equal the graph of the map. Give a geometric interpretation of the additional components in the case where the X-variety strictly contains the graph.

⋆**18.** Suppose that p/q is a proper rational map between balls. For a fixed allowable denominator, describe the geometry of the set of allowable numerators. Note that, even when the denominator is a constant, the problem is not trivial. Proposition 5.1 gives a complicated (necessary and sufficient) condition on the numerator.

⋆*19.* Prove a result about boundary behavior of the Kobayashi metric on a domain of finite type that includes information about the associated family of ideals.

⋆*20.* Prove or disprove the conjecture in Chapter 6, Section 3, concerning precise subelliptic estimates.

⋆*21.* Generalize the notion of subelliptic estimate to take into account the associated ideals of holomorphic functions.

Index of Notation

$\mathbb{C}^n$	Complex Euclidean space of dimension n.
$\mathbb{R}^n$	Real Euclidean space of dimension n.
$\mathbb{Z}$	The integers.
$f : \Omega_1 \to \Omega_2$	Notation for a function with given domain and range.
$\Omega \subset\subset \mathbb{C}^n$	Notation meaning that Ω has compact closure in $\mathbb{C}^n$.
$\frac{\partial}{\partial z^j}, \frac{\partial}{\partial \bar{z}^j}, d, \partial, \bar{\partial}$	Fundamental partial derivative operators, the exterior derivative, etc.
$\partial_o P$	The distinguished boundary of a polydisc.
$b\Omega$	The boundary of an open set.
$O(\Omega), C(\Omega)$	Spaces of holomorphic and continuous functions.
$_n O_p = O$	The ring of germs of holomorphic functions.
$O[w]$	The polynomial ring with coefficients in O.
$\mathbf{V}(I)$	The variety of an ideal.
$\mathrm{rad}(I)$	The radical of an ideal.
$\mathbf{I}(V)$	The ideal of a variety.
M	The maximal ideal in O.
K_f	The Bochner–Martinelli kernel.
$\mathbf{D}(f), \mathbf{D}(I), \mathbf{D_q}(I)$	Codimension of a mapping, ideal, and qth order generalization.
$\mathbf{T}(I)$	Order of contact of an ideal.
$v(z)$	Multiplicity of a parameterized curve at the origin.
$\mathbf{K}(I)$	Smallest exponent for an M-primary ideal.
$\mathbf{Res}(f)$	The Grothendieck residue of a mapping.
$\mathrm{index}_0(f)$	The index (topological degree) at 0 of a mapping.
$T^{1,0}(M), T_p^{1,0}(M), T^{0,1}M, HM, hM$	Bundles associated with a CR manifold.
$\lambda\left(L, \overline{K}\right) = \left\langle \eta, \left[L, \overline{K}\right]\right\rangle$	The Levi form.

$\mathbf{U}(N)$	The unitary group on $\mathbb{C}^n$.
$I(U,p)$	The associated ideals.
$\Delta^1(M,p)$	The one-type of a point on a real subvariety.
$\Delta^q_{\mathrm{reg}}(M,p)$	The regular q-type of a point.
$\Delta^q(M,p)$	The q-type of a point.
$B^q(M,p)$	The multiplicity of a point on a real hypersurface.
$j_k r = j_{k,p} r$	The kth order Taylor polynomial of r.
$[L,B]$	The Lie bracket of two vector fields.
$\langle \eta, L \rangle$	Contraction of a 1-form and a vector field.
$\langle \omega, L \wedge B \rangle$	Contraction of a 2-form and two vector fields.
$\mathrm{type}_p(L)$	The type of a vector field (number of commutators).
$c_p(L)$	A measurement of the order of vanishing of the Levi form on a vector field.
B_n	The unit ball in $\mathbb{C}^n$.
$\mathrm{Aut}(B_n)$	The automorphism group of the unit ball.
ξ_a	A certain linear fractional transformation moving the origin to a.
$L_a(z) = sz + \frac{\langle z,a \rangle}{s+1} a$	A certain linear map.
$f \in C^\infty(\overline{\Omega})$	Notation for f being infinitely differentiable on the closed domain.
C^k	Having k continuous derivatives.
$g \otimes h = (g_1 h_1, g_2 h_1, \ldots,$ $g_n h_1, g_1 h_2, \ldots, g_n h_N)$	The tensor product of two mappings.
$\mathbb{C}^N = A \oplus A^\perp$	An orthogonal decomposition.
$f = f_A \oplus f_{A^\perp}$	The induced orthogonal decomposition on a function.
$E(A,g)(f) = (f_A \otimes g) \oplus f_{A^\perp}$	The extend operation.
$E^{-1}(A,g)((f_A \otimes g) + f_{A^\perp}) = f$	The inverse of the extend operation.
$V = V(n,m)$	The space of polynomials of degree at most m in n variables.
$H_m = z \otimes z \ldots \otimes z$	The m-fold tensor product.
$\pi : G \to \mathbf{U}(n)$	A unitary representation of a finite group.
$\Gamma(p,q)$	A certain fixed-point–free finite unitary group.
$L(p,q)$	A Lens space.

$S^3/\Gamma\,(p,q)$	Definition of $L\,(p,q)$.
$v_2\,(p)$	The number of factors of two in an integer.
$\Upsilon\,(z,\overline{z}) = -\prod_{\gamma\in\Gamma}\left(1 - \langle\gamma z, z\rangle\right)$	The minimal invariant polynomial.
$\|\phi\|_\psi^2 = \int_\Omega \sum_J |\phi_J|^2\, e^{-\psi}\, dV$	The weighted L^2 norm of a differential form.
Λ^s	The standard tangential pseudodifferential operator of order s.
$\|\|f\|\|_s^2 = \int_{-\infty}^0 \int_{R^{2n-1}} |\Lambda^s f\,(t,r)|^2\, dt\, dr$	The squared tangential Sobolev s-norm.
$\overline{\partial}^*$	The adjoint of $\overline{\partial}$.
$\Box = \left(\overline{\partial}\,\overline{\partial}^* + \overline{\partial}^*\overline{\partial}\right)$	The complex Laplacian.
$\mathrm{Ker}\,(\Box)$	The kernel of the complex Laplacian.
$N = \Box^{-1}$ on $(\mathrm{Ker}\,(\Box))^\perp$	The $\overline{\partial}$-Neumann operator.
$L^2\,(\Omega)$	Square integrable functions.
$\mathbf{L}\left(A^{0,1}\right)$	Linear transformations on the space of $(0,1)$ forms.
$P : L^2\,(\Omega) \to L^2\,(\Omega) \cap O\,(\Omega)$	The Bergman projection.

Bibliography

[Ah] Ahlfors, L. V., *Complex Analysis*, McGraw-Hill, New York, 1966.

[Al] Alexander, H., Proper holomorphic mappings in $\mathbb{C}^n$, *Indiana Univ. Math. J.*, 26 (1977), 137–146.

[AGV] Arnold, V. I., Gusein-Zade, S. M., and Varchenko, A. N., *Singularities of Differentiable Maps*, Volume I, Birkhauser, Boston, 1985.

[BR1] Baouendi, M. S., and Rothschild, L. P., Germs of CR maps between real analytic hypersurfaces, *Inventiones Mathematicae*, 93 (1988), 481–500.

[BR2] Baouendi, M. S., and Rothschild, L. P., Geometric properties of mappings between hypersurfaces in complex space, *J. Diff. Geom.*, 31 (1990), 473–499.

[BR3] Baouendi, M. S., and Rothschild, L. P., Holomorphic mappings of real analytic hypersurfaces. In *Proceedings of Symposia in Pure Mathematics*, Vol. 52, Part 3, Several Complex Variables and Complex Geometry, American Mathematical Society, Providence, RI, 1991, 15–26.

[BJT] Baouendi, M. S., Jacobowitz, H., and Treves, F., On the analyticity of CR mappings, *Ann. Math.*, 122 (1985), 365–400.

[BT] Baouendi, M. S., and Treves, F. , About the holomorphic extension of CR functions on real hypersurfaces in complex space, *Duke Math. J.*, 51 (1984), 77–107.

[Ba1] Barrett, D., Boundary singularities of biholomorphic maps. In *Lecture Notes in Mathematics* 1268, pp. 24–28, Springer-Verlag, Berlin, 1987.

[Ba2] Barrett, D., Biholomorphic domains with inequivalent boundaries, *Inventiones Math.*, 85 (1986), 373–377.

[Ba3] Barrett, D., Irregularity of the Bergman projection on a smooth bounded domain in $\mathbb{C}^2$, *Ann. Math.*, 119 (1984), 431–436.

[BFG] Beals, M., Fefferman, C., and Grossman, R., Strictly pseudoconvex domains in $\mathbb{C}^n$, *Bulletin AMS*, 8(2) (1983), 125–322.

[Bed] Bedford, E., Proper holomorphic maps, *Bulletin AMS*, 10(2) (1984), 157–175.

[BB]   Bedford, E., and Bell, S., Proper self maps of weakly pseudoconvex domains, *Math. Annalen*, 261 (1982), 47–49.

[Be1]   Bell, S., Biholomorphic mappings and the $\bar{\partial}$ problem, *Ann. Math.*, 114 (1981), 103–113.

[Be2]   Bell, S., An extension of Alexander's theorem on proper self mappings of the ball in $\mathbb{C}^n$, *Indiana Univ. Math. J.*, 32 (1983), 69–71.

[Be3]   Bell, S., Mapping problems in complex analysis and the $\bar{\partial}$ problem, *Bulletin AMS*, 22(2) 1990, 233–259.

[BC1]   Bell, S., and Catlin, D., Boundary regularity of proper holomorphic mappings, *Duke Math. J.*, 49 (1982), 385–396.

[BC2]   Bell, S., and Catlin, D., Regularity of CR mappings, *Math. Z.*, 199 (1988), 357–368.

[Bl]   Bloom, T., On the contact between complex manifolds and real hypersurfaces in $\mathbb{C}^3$, *Trans. AMS*, 263 (1981), 515–529.

[BlG]   Bloom, T., and Graham, I., A geometric characterization of points of type m on real submanifolds of $\mathbb{C}^n$, *J. Diff. Geom.*, 12 (1977), 171–182.

[Bo]   Boas, H. P., Small sets of infinite type are benign for the $\bar{\partial}$-Neumann problem, *Proc. AMS*, 103(2) (1988), 569–577.

[BS1]   Boas, H. P., and Straube, E. J., Equivalence of regularity for the Bergman projection and the $\bar{\partial}$-Neumann operator, *Manuscripta Math.*, 67 (1990), 25–33.

[BS2]   Boas, H. P., and Straube, E. J., Sobolev estimates for the $\bar{\partial}$-Neumann operator on domains admitting a defining function that is plurisubharmonic on the boundary, *Math. Z.*, 206 (1991), 81–88.

[Bg]   Boggess, A., *CR Manifolds and the Tangential Cauchy Riemann Complex*, CRC Press, Boca Raton, FL, 1991.

[Bk]   Bourbaki, N., *Commutative Algebra, Elements of Mathematics*, Springer-Verlag, Berlin, 1989.

[C1]   Catlin, D., Necessary conditions for subellipticity of the $\bar{\partial}$-Neumann problem, *Ann. Math.*, 117 (1983), 147–171.

[C2]   Catlin, D., Boundary invariants of pseudoconvex domains, *Ann. Math.*, 120 (1984), 529–586.

[C3]   Catlin, D., Subelliptic estimates for the $\bar{\partial}$-Neumann problem on pseudoconvex domains, *Ann. Math.*, 126 (1987), 131–191.

[Ch]   Chiappari, S., Proper holomorphic mappings of positive codimension in several complex variables. Doctoral thesis, University of Illinois, Urbana, Illinois (1990).

[CS1]   Cima J., and Suffridge, T., A reflection principle with applications to proper holomorphic mappings, *Math. Annalen*, 265 (1983), 489–500.

[CS2]   Cima J., and Suffridge, T., Proper mappings between balls in $\mathbb{C}^n$, *Lecture Notes in Mathematics* 1268, 66–82, Springer-Verlag, Berlin, 1987.

[CS3]   Cima J., and Suffridge, T., Boundary behavior of rational proper maps, *Duke Math J.*, 60(1) (1990), 135–138.

[D1]   D'Angelo, J., Real hypersurfaces, orders of contact, and applications, *Ann. Math.*, 115 (1982), 615–637.

[D2]   D'Angelo, J., A Bezout type theorem for real hypersurfaces, *Duke Math J.*, 50 (1983), 197–201.

[D3]   D'Angelo, J., Intersection theory and the $\bar{\partial}$-Neumann problem, *Proc. Symp. Pure Math.*, 41 (1984), 51–58.

[D4]   D'Angelo, J., Finite type conditions for real hypersurfaces, *Lecture Notes in Mathematics* 1268, pp. 83–102, Springer, Berlin, 1987.

[D5]   D'Angelo, J., Iterated commutators and derivatives of the Levi form, *Lecture Notes in Mathematics* 1268, pp. 103–110, Springer, Berlin, 1987.

[D6]   D'Angelo, J., Proper holomorphic maps between balls of different dimensions, *Michigan Math. J.*, 35 (1988), 83–90.

[D7]   D'Angelo, J., The notion of formal essential finiteness for smooth real hypersurfaces, *Indiana Univ. Math. J.*, 36 (1987), 897–903.

[D8]   D'Angelo, J., Proper polynomial mappings between balls, *Duke Math J.*, 57 (1988), 211–219.

[D9]   D'Angelo, J., Polynomial proper holomorphic mappings between balls, II, *Michigan Math. J.*, 38 (1991), 53–65.

[D10]   D'Angelo, J., Finite type and the intersections of real and complex varieties, In *Proceedings of Symposia in Pure Mathematics*, Vol. 52, Part 3, Several Complex Variables and Complex Geometry, American Mathematical Society, Providence, RI, 1991, 103–117.

[D11]   D'Angelo, J., The geometry of proper holomorphic maps between balls, *Contemporary Mathematics*, Vol. 137 (1992), 191–215.

[DL]   D'Angelo, J., and Lichtblau, D. A., Spherical space forms, CR maps, and proper maps between balls, *J. Geom. Anal.*, 2 (1992) 391–415.

[DF1]   Diederich K., and Fornaess, J. E., Pseudoconvex domains with real analytic boundary, *Ann. Math.*, 107(2) (1978), 371–384.

[DF2]   Diederich K., and Fornaess, J. E., Boundary regularity of proper holomorphic mappings, *Inventiones Math.*, 67 (1982), 363–384.

[DW]   Diederich, K., and Webster, S., A reflection principle for degenerate real hypersurfaces, *Duke Math. J.*, 47 (1980), 835–843.

[Dr]   Dor, A., Proper holomorphic maps between balls in one codimension, *Arkiv för matematik*, 28(1) (1990), 49–100.

[EL]   Eisenbud, D., and Levine, H., An algebraic formula for the degree of a C^∞ map germ, *Ann. Math.*, 106 (1977), 19–44.

[Fa1]   Faran, J., Maps from the two-ball to the three ball, *Inventiones Math.*, 68 (1982), 441–475.

[Fa2]   Faran, J., On the linearity of proper maps between balls in the low codimension case, *J. Diff. Geom.*, 24 (1986), 15–17.

[Fe]   Fefferman, C., The Bergman kernel and biholomorphic mappings of pseudoconvex domains, *Inventiones Math.*, 26 (1974), 1–65.

[FK]   Folland, G., and Kohn, J., The Neumann problem for the Cauchy–Riemann complex, *Ann. Math. Studies*, 75, Princeton Univ. Press, 1972.

[Fo1]   Forstnerič, F., Proper holomorphic maps from balls, *Duke Math. J.*, 53(2) (1986), 427–441.

[Fo2]   Forstnerič, F., Proper holomorphic maps: A survey. In *Proceedings of the Mittag-Leffler Institute, 1987–88*, Princeton University Press, Princeton, NJ, to appear.

[Fo3]   Forstnerič, F., Extending proper holomorphic mappings of positive codimension, *Inventiones Math.*, 95 (1989), 31–62.

[Fo4]   Forstnerič, F., Embedding strictly pseudoconvex domains into balls, *Trans. AMS*, 295(1) (1986), 347–368.

[Fu]   Fulton, W., *Intersection Theory*, Springer-Verlag, Berlin, 1984.

[GR]   Grauert, H., and Remmert, R., *Coherent Analytic Sheaves*, Springer-Verlag, New York, 1984.

[Gn]   Greiner, P., Subelliptic estimates of the $\bar\partial$-Neumann problem in $\mathbb{C}^2$, *J. Diff. Geom.*, 9 (1974), 239–260.

[GS]   Greiner, P., and Stein, E., *Estimates for the $\bar\partial$-Neumann Problem*, Princeton University Press, Princeton, NJ, 1977.

[GH]   Griffiths, P., and Harris, J., *Principles of Algebraic Geometry*, Wiley, New York, 1978.

[Gu1]   Gunning, R., *Lectures on Complex Analytic Varieties: The Local Parameterization Theorem*, Princeton University Press, Princeton, NJ, 1970.

[Gu2]   Gunning, R., *Lectures on Complex Analytic Varieties: Finite Analytic Mappings*, Princeton University Press, Princeton, NJ, 1974.

[Ha]   Hakim, M., Applications holomorphes propres continues de domaines strictement pseudoconvexes de $\mathbb{C}^n$ dans les boule unite de $\mathbb{C}^{n+1}$, *Duke Math. J.*, 60 (1990), 115–134.

[Hm]   Hörmander, L., *An Introduction to Complex Analysis in Several Variables*, Von Nostrand, Princeton, NJ, 1966.

[Ki]   Kiselman, C., A study of the Bergman projection in certain Hartogs domains. In *Proceedings of Symposia in Pure Mathematics*, Vol. 52, Part 3, Several Complex Variables and Complex Geometry, American Mathematical Society, Providence, RI, 1991, 219–231.

[K1] Kohn, J., Harmonic integrals on strongly pseudoconvex manifolds I, II, *Ann. Math.*, 2 (1963), 112–148; ibid. 79 (1964), 450–472.

[K2] Kohn, J., Global regularity for $\bar{\partial}$ on weakly pseudoconvex manifolds, *Trans. AMS*, 181 (1973), 273–292.

[K3] Kohn, J., A survey of the $\bar{\partial}$-Neumann problem. In *Proceedings of the Symposia on Pure Mathematics*, Providence, RI, 1984, 137–145.

[K4] Kohn, J., Subellipticity of the $\bar{\partial}$-Neumann problem on pseudoconvex domains: Sufficient conditions, *Acta Math.*, 142 (1979), 79–122.

[Kr] Krantz, S., *Function Theory of Several Complex Variables*, Wiley, New York, 1982.

[La] Lang, S., *Introduction to Differentiable Manifolds*, Wiley, New York, 1962.

[Le1] Lempert, L., Imbedding Cauchy-Riemann manifolds into a sphere, *Int. J. Math.*, 1 (1990), 91–108.

[Le2] Lempert, L., On the boundary behavior of holomorphic mappings. In *Contributions to Several Complex Variables in honor of Wilhelm Stoll*, A. Howard and P. M. Wong, eds., pp. 193–215, Friedr. Vieweg & Sons, Braunschweig/Weisbaden, 1986.

[Li1] Lichtblau, D., Invariant proper holomorphic maps between balls. Doctoral thesis, University of Illinois, Urbana, Illinois (1991).

[Li2] Lichtblau, D., Invariant proper holomorphic maps between balls, *Indiana Univ. Math. J.*, 41(1) (1992), 213–231.

[Lw] Løw, E., Embeddings and proper holomorphic maps of strictly pseudoconvex domains into polydisc and balls, *Math. Z.*, 190 (1985), 401–410.

[M] Malgrange, B., *Ideals of Differentiable Functions*, Oxford University Press, London, 1966.

[Mc1] McNeal, J., Convex domains of finite type, *J. Functional Anal.*, 108 (1992), 361–373.

[Mc2] McNeal, J., Estimates for the Bergman kernel on convex domains (preprint).

[N] Narasimhan, R., *Introduction to the Theory of Analytic Spaces*, Lecture Notes 25, Springer-Verlag, New York, 1966.

[P1] Pinchuk, S., On the analytic continuation of holomorphic mappings, *Math USSR-Sb.*, 27 (1975), 375–392.

[P2] Pinchuk, S., Holomorphic maps in $\mathbb{C}^n$ and the problem of holomorphic equivalence. In *Encyclopedia of Math. Sciences*, Vol. 9, G. M. Khenkin, ed., Several Complex Variables III, 173–200. Springer-Verlag, Berlin, 1980.

[Ra]   Range, R. M., *Holomorphic Functions and Integral Representation Formulas in Several Complex Variables*, Springer-Verlag, New York, 1986.

[RS]   Remmert, R., and Stein, K., Eigentliche holomorphe abbildungen, *Math. Z.*, 73 (1960), 159–189.

[Ro1] Rosay, J. P., Injective holomorphic mappings, *Amer. Math. Monthly*, 89 (1982), 587–588.

[Ro2] Rosay, J. P., New examples of non-locally embeddable CR structures, *Ann. Inst. Fourier* (Grenoble), 39 (1989), 811–823.

[Ru1] Rudin, W., *Function Theory in the Unit Ball of* $\mathbb{C}^n$, Springer-Verlag, New York, 1980.

[Ru2] Rudin, W., Homogeneous proper maps, *Nederl. Akad. Wetensch. Indag. Math.*, 46 (1984), 55–61.

[Ru3] Rudin, W., Proper holomorphic maps and finite reflection groups, *Indiana Univ. Math. J.*, 31(5) (1982), 701–719.

[Sfv] Shafarevich, I., *Basic Algebraic Geometry*, Springer, Berlin, 1977.

[Shi] Shiffman, B., Separate analyticity and Hartogs theorems, *Indiana Univ. Math. J.*, 38(4) (1989), 943–957.

[Si1] Sibony, N., Une classe de domaines pseudoconvexes, *Duke Math. J.*, 55 (1987), 299–319.

[Si2] Sibony, N. Some aspects of weakly pseudoconvex domains. In *Proceedings of Symposia in Pure Mathematics*, Vol. 52, Part 1, Several Complex Variables and Complex Geometry, American Mathematical Society, Providence, RI, 1991, 199–232.

[T]   Trepreau, J. M., Sur le prolongement holomorphe des fonctions CR definies sur une hypersurface reele de classe C^2 dans $\mathbf{C}^n$, *Inventiones Math*, 83 (1986), 583–592.

[V]   Van der Waerden, B. L., *Modern Algebra*, two volumes, Frederick Ungar Publishing Co., New York, 1949.

[We1] Webster, S., On mapping an n ball into an $n+1$ ball in complex space, *Pacific J. Math.*, 81 (1979), 267–272.

[We2] Webster, S., On the mapping problem for algebraic real hypersurfaces, *Inventiones Math.*, 43 (1977), 53–68.

[We]   Wells, R. O., Jr., *Differential Analysis on Complex Manifolds*, Prentice-Hall, Englewood Cliffs, NJ, 1973.

[Wh]   Whitney, H., *Complex Analytic Varieties*, Addison-Wesley, Reading, MA, 1972.

[Wo]   Wolf, J., *Spaces of Constant Curvature*, McGraw-Hill, New York, 1967.

Index